# INVESTING IN
# **A NEW CLIMATE**

## A SUSTAINABLE APPROACH
## TO INVESTING & LIVING IN A NEW CLIMATE

Scott Schwartz

© 2021 Scott Schwartz

All rights reserved.

Library of Congress Control Number: 2021911775

ISBN: 978-1-7373329-0-9

Contributor: John Boyer

Cover and book design: Jill Turney Design

Cover photos:
Banknotes © Miriam Doerr, Dreamstime; View of Lyon, France at sunrise © Hornet83, Dreamstime; Shanghai Bund skyline landmark at Ecological energy solar panel © Ipadimages, Dreamstime

This book is dedicated to Jessica and Danielle.

**Looking at the world through a sustainability lens not only helps us 'future proof' our supply chain, it also fuels innovation and drives brand growth.**

**—Paul Polman**

(CEO) of the British-Dutch consumer goods company Unilever. In 2019, he created a new organization called Imagine, to help businesses "eradicate poverty and inequality and stem runaway climate change."

# Table of Contents

# FOREWORD

So many books have been written about how, where, and when to invest that if stacked one upon the other, the column would likely reach halfway to the moon. I have read quite a few of them and, other than Graham and Dodd's classic on value investing, have found them poorly written and uninteresting.

Here, however, is a rare example of an investment book that is an enjoyable read and where the author has done an outstanding job of capturing the modern investment approach to putting one's money into sectors that reflect social responsibility and personal interest.

I have known Scott for several years. Our experiences were similar in many ways, from adventure travel to sailing and motorcycle riding. The main theme that really brought us together was interest in the outdoors and our observations of the changing climate and the human impact on it. There is nothing so attention-getting as seeing changes for oneself . My Silent Spring moment was growing up in the Virginia countryside. There was a boggy area on our property that was a haven for hundreds of butterflies, from swallowtails to monarchs. One day they were gone, never to return, due to the overuse of DDT. It was a jolting moment for a boy of ten and had a deep impact on my way of looking at how people interact with the natural world.

Today, we face multiple problems with the way societies treat the environment. Global warming is certainly one of the key issues we face, but there are others. Overfishing, slash and burn agricultural farming, unsustainable livestock management methods and over consumption are but a few.

Scott has put together a fascinating book that leads the reader through the history of global expansion, trade and the crucial interactions between human endeavor and our limited natural resources, whether they be water or power generation. Added to these are examples of how investors can make informed decisions about which companies represent opportunities to profit from informed policy and product development and how to develop a rational ESG and SRI portfolio.

Peter Craddock
Founder
Shoreline Venture Management

# PREFACE

## Adapting to a Changing Climate

This is not a book about doom and gloom, but about adapting to change. Climate change is about the shifting of circumstances we have long taken for granted. To survive — and succeed— we must adapt. We must learn how to live and how to plan ahead by making reasonable extrapolations of circumstances within a changing global climate. Warming of the climate system is unequivocal and, since the 1950s, many of the observed changes are unprecedented over decades to millennia. The atmosphere and oceans have warmed, amounts of snow and ice have diminished, sea level has risen, and the concentration of greenhouse gases has increased.

Even the naysayers, such as the authors of the book, *Climate of Extremes: Global Warming Science They Don't Want You to Know*, discuss scientific flaws and exaggerations by some scientists, but they do not outright deny the climate is changing (Michaels & Balling, 2010). The authors primarily argue the degree of change and whether the change is anthropogenic. But they do not deny that climate change is happening.

*Investing in a New Climate* is about dealing with a new normal. This new normal has to do with a changing climate, but not solely the meteorological climate. The concept of the changing climate also has to do with what a new meteorological climate means to economies, industries, and financial markets around the world. The meteorological climate is one catalyst, but I am talking about Climate with a big "C." Adapting to the changing climate also includes advances in technology such as in medicine, energy, transportation, space travel, artificial intelligence, and computing.

This book is not a debate over what is causing climate change: there are plenty of books that discuss that. This book is an approach to dealing with a changing climate. Climate change is inevitable, a static climate is short-lived. There is no doubt that the Earth is getting warmer. We can put our head in the sand and not accept that change is taking place, or we can take a proactive and sensible approach to dealing with the changing climate.

What can we do? First, get a handle on the facts of what's happening. Second, determine what we can do about it in our own lives. And finally, put together a plan to help mitigate changes and to adapt to the inevitable changes. We can work to understand economic trends, look at industrial impacts and modify our lifestyles. We need to analyze our current investments and consider updating our investment portfolios. We might consider modifying our vacation plans, such as choosing to go to that exotic locale now before it disappears underwater or becomes too dangerous to visit. With sea levels rising, will Tuvalu, parts of Fiji, Kiribati, and The Maldives survive? In the U.S., are Boston, Baltimore, Miami, New York, Philadelphia, Providence, Rhode Island and the San Francisco Bay Area at risk? What are the ramifications of sea level rise, who will benefit, who will suffer? Should we invest in northern European agriculture industries, companies that build dikes, and/or water purification technologies? Perhaps we should not retire to a seashore that is about to be underwater.

Meteorological changes affect economic trends. As the Earth gets warmer, northern latitudes will become more temperate, while equatorial and sub-equatorial climates could become more stressed. Countries in higher latitudes that are generally more economically successful will likely weather the change better, while lower latitude countries that are currently on the edge will likely become even more stressed. The approach of this book is to discuss the history of climate change, the science of what has happened in the past, what is happening now, and what will likely happen in the future. Most importantly, how can we mitigate negative effects of climate change, adapt to changing circumstances, and thrive as the world changes around us?

With so many types of climate change occurring (economic, meteorological, industrial), we don't have the luxury of denying that change is happening. There is a metaphoric thunderstorm in front of our planet. As a professional pilot, I have faced quite a few storms. In flight, we have three options: fly around the storm, turn around, or fly through the storm. Most of us don't want to turn around or fly through the storm. Most of us think modifying our course and avoiding the storm is the most prudent approach. So, let's gather the facts, analyze the situation, and modify our course. And let's look at the troubling, new, and exciting opportunities that emerge.

# INTRODUCTION

I have always been a student of weather and climate. Not just from my armchair, but observing it firsthand, in the real world. I spent my childhood exploring the High Sierra, then I took to the skies and built a career as a pilot. I have sailed around the world and spent time climbing and trekking on multiple continents. It is my personal observation that the climate is changing in a big way.

In many ways, I have been working a lifetime to create this book. As a kid, I was always interested in science and economics. After spending a lot of time outdoors and camping in the wilderness of California and Nevada, I realized early that there was a strong intrinsic connection between people and nature. Now I see that translating into a strong connection between the climate and financial markets.

In 5th grade, I learned about speculating in the stock market. I was shocked that one could make money by buying and selling ownership in a company. How cool, I thought, to make money without working. Well, that was a flawed assumption. In my present view, ditch digging is a more consistent method of making money than gambling in the financial markets, which isn't to say that thoughtful strategic and tactical investing isn't a good idea. After decades working in the financial services business, I believe the financial markets provide great tools for building wealth and offsetting risk.

I grew up in San Francisco in the late 1960s and early 1970s. It was my maternal uncle Mort who first introduced me to aviation. He worked for Pan American World Airlines as a dispatcher, based in Beirut. He peppered holiday dinners with great stories of daring feats and close calls of near and real aircraft disasters. I also recall a decommissioned military jet fighter aircraft on the Larsen Park playground, on 19th Avenue in the Sunset District of San Francisco. It was effectively a playground toy, though it was a real airplane. It's gone today, but I spent many hours pretending to fly that jet in the summers. About seven years later, at the tender age of 16, I took skydiving lessons in the nearby community of Antioch, California, jumping out of sketchy, 1960s vintage single-engine Cessna planes. "Why would anybody want to jump out of a perfectly good airplane?"—that did not apply in this case, as the planes had no cabin doors, no seats, little paint, and lots of dings and scratches. We sat on the floor of the plane as we slowly ascended to jump altitude. (Since we regularly exceeded the recommended gross weight for the aircraft, our climb seemed to take forever.) I was always slightly relieved after jumping out of the plane, no longer in danger. It's a wonder there weren't more accidents.

After a couple of minor accidents — landing injuries, really — I decided to focus on becoming an airline pilot. My flying career has included destinations from the urban centers of the East and West Coasts, the Plains States, and the South. I have enjoyed beignets in New Orleans, seen superstorms in Oklahoma, experienced three hurricanes in Florida, and watched many sunrises in Hawaii. I also had the added excitement of experiencing erupting volcanoes. I have been huddled up in the bitter cold of Calgary, Canada, suffered torrential rain in Toronto, and seen the Northern Lights. I have been amazed by the expanse of the unpopulated parts of the United States, Mexico, and Canada, especially Newfoundland and Labrador and the Maritime provinces of Eastern Canada.

In the mid-1990s, the airline industry was going through a tough time, so I thought I'd hedge my career bets and pursue my other passion, finance and economics. I went to graduate school at City University in Washington State and received an MBA with an emphasis in finance and economics. After that, I did some post-graduate work in finance and investments through programs at The Wharton School, The University of Pennsylvania, and through UC Berkeley's financial planning program to further diversify my education in the world of finance. Over the years, I've attended professional financial workshops and continuing education at Stanford University, Harvard Business School, Oxford, and EDHEC in Paris. I started out as a part-time stockbroker and ended up as an investment advisor for an investment bank and was associated with a few investment firms before I started my own company— still flying of course, though less often. As an investment advisor, I have worked with individuals and companies, and consulted on retirement plans for Silicon Valley tech companies. I also oversaw several airline pension plans.

Besides a lengthy career in the air and in the financial industry, I have also been strongly drawn to ocean sailing. I've had the privilege of crewing on sailboats since I was a teenager. I learned to sail from fellow pilots who had access to sailboats. I have crewed on and led several ocean-going sailing vessels, both cruising and racing. I earned a U.S. Coast Guard Master's license, which allowed me to sail as captain or crew for hire.

Through the air, sea, and land, I've had the opportunity to explore all seven continents, much of it on foot. Some of the exploration included climbing Tanzania's Kilimanjaro, Japan's Mt. Fuji, Hawaii's Mauna Kea, Washington's Mt. Rainer, California's Mt. Shasta, and places in the Sierra range, the Himalayas, and in Antarctica. One of the common threads of these enterprises was dealing with the weather. Before each flight, sail, or trek, I studied the weather and interacted with various meteorological experts: professional dispatchers and meteorologists with the airlines, weather routers in the marine world, and the National Weather Service, local rangers, or their foreign equivalents when climbing or trekking. It was critical to have up-to-date information on the weather, as our interactions with weather could have life and death consequences.

My experiences and activities were not particularly unique, and definitely not heroic. I am not claiming to be an expert, but I do have the benefit of firsthand experience and a three-dimensional global perspective from the air, land, and sea. The climate is changing. It's getting warmer, storms are getting more severe, and droughts are becoming worse and more frequent. The effects of the changing climate are pervasive and directly impact public health in several ways: storms and rising sea levels cause displacement of populations and contamination of water sources that can result in disease; increased surface temperatures have a profound impact on air quality, both directly and through erosion and fires, and can cause or exacerbate respiratory illnesses; drought can result in malnutrition and famine. All of these changes can impact our physical and mental health. Everywhere we turn, climate change is affecting our lives. Based on my own education and experiences and after talking with various experts I ran across during my travels, I reached the conclusion that climate change deserves further study. I decided to dig deeper into the disruptive impacts of the changing climate and how those changes will likely affect the global economy now, and for centuries to come.

The climate is changing. Not only the physical climate, but the social, economic, and psychological climate, as well. We are living in a disrupted world that is increasingly divided along the lines of economics and politics. Already, globally, we think, feel, and act differently than those before us did. This book is about adaptation and how to take advantage of and thrive in a changing world. How can we learn to gauge predictable climate change and interpret it as an opportunity to quantify probable impacts and act on them to mitigate economic damage and benefit from new opportunities? While not a crystal ball, history holds clues of what may likely come, as well as examples of how humans have adapted to changing environments. Through the lens of scientific objectivity, history can give us insight into what changes in our meteorological and economic climates we can anticipate so that we can adapt and find success.

✗ ✗ ✗ ✗ ✗ ✗ ✗ ✗

I have always been curious and adventurous. I became an airline pilot in part because I thought that the lifestyle of Frank Abagnale as depicted in the movie If You Can suited me; the idea of an office that moved at 600 mph (about 80% of the speed of sound) seemed appealing. After preliminary civilian flight training and a couple years of college focusing on aviation flight science and aircraft maintenance, I had to declare a major to finish my bachelor's degree. I was torn between meteorology and astrophysics. I thought maybe I could leverage my interest and experience flying into the Astronaut Program, but with cuts at NASA and the highly competitive nature of the Space Program, I decided to go for the path that seemed more likely for success. So off I went to Weather and Climate school through the Geography Department at San Francisco State University. At the time, the airlines were not interested in what a pilot studied in college, as long as we had a four-year degree and a reasonable GPA. So, I majored in something that I had a serious interest in and could also be helpful as preparation to becoming a professional pilot. Pilots do consider weather as part of their flight planning process on a continual basis, one would hope.

I started out as an instructor at the local flight school in San Carlos, California. I quickly graduated to flying charters for the local TV station news department, Pacific Gas and Electric (PG&E), the British Broadcasting Corporation (BBC), and for the U.S. Geological Survey (USGS). A couple of years later in Marin County, California, I flew corporate planes for a West Coast music promoter, Bill Graham, and the rock bands he promoted. At the time, Graham's bands included the Grateful Dead, The Jefferson Starship, Santana, and Huey Lewis and the News. I also flew for Lucas Films, including flying George Lucas and his then-girlfriend Linda Ronstadt, as well as doing aerial set-up shots for Lucas's film, Indiana Jones and the Temple of Doom.

In 1985, at the age of 25, I ended up as a pilot based in New York at Trans World Airlines (TWA). I flew around Europe, the Middle East, the Caribbean, and many locations in the U.S. Like most pilots, I was promoted through the ranks from Flight Engineer to Co-Pilot, and finally Captain. Ultimately, TWA's assets were purchased by American Airlines and I became an American Airlines pilot. Through my background in finance, and as a member of the ad hoc pilot fraternity, I was asked to work with the TWA pilot's association known as ALPA (Air Line Pilots Association) and later for the American Airlines pilots' group, APA (Allied Pilots Association). Among other things, I was involved in oversight of pensions, 401k plans, and Mergers and Acquisitions as a representative for the pilot group. I was asked to represent the pilots through four corporate bankruptcies, several potential and actual mergers, and I even did a stint on the Unsecured Creditors Committee during the American Airlines restructure of 2013. During many of those years I rarely flew, but I gained up-close and personal experience with corporate finance. I was privy to lots of non-public information, and a lot of it was not pretty. Billions of dollars were at stake, and tens of thousands of jobs were at risk.

Once I got back in the air, I flew extensively south of the U.S. border to Mexico, Belize, and the Caribbean. These areas were both tropical as well as arid and dry. I enjoyed both sweltering heat and oppressive humidity with occasional bouts of perfect sun drenched, not too hot or sticky, weather. I have experienced most of the major cities in Latin America: I particularly enjoyed Mexico City, with its rich culture and endless extremes of wealth, poverty, history, and cultural diversity. From Puerto Rico to just north of Venezuela, the Caribbean is a tapestry of idyllic scenery, abject poverty, luxury, and violence, highlighted with devastating hurricanes. In the late 1980s through the late 1990s, I spent time flying to and from and within Europe. This included a brief stint at TWA's Berlin base. TWA had a base in Berlin as result of the Allied Powers Act which was post World War II agreement giving former "allies" special rights to operating airliners in Germany.

Europe is a complicated and diverse place offering a range of intolerable summer heat in the Mediterranean to the bone chilling brutal cold of Central Europe in the winter. I spent years exploring cafes in Paris, churches in Rome, pubs in London, coffee houses in Brussels, and seafood restaurants in Lisbon. I have also observed the rising waters of Venice and the Netherlands. From the air, I saw the Middle East as a dry place, a far cry from the Fertile Crescent of millennia ago. Having flown to both Egypt and Israel, I noted that, while geographically close, they are very different in terms of land use and agriculture.

I found that my time in Weather and Climate school was beneficial. When pilots execute a flight, we input an estimated outside air temperature at cruise altitude into our inflight computer. In the mid-1980s, the temperature was normally a few degrees cooler or warmer than expected; since the mid-1990s, I have observed the temperature is almost always warmer than expected, almost never colder, and often warmer by over 10° Celsius. There's a well-known tipping point of 2° Celsius at sea level that is often discussed during climate change talks, but at higher altitudes I have observed a greater than 2° increase. I'm not claiming extraordinary significance to this other than my personal and non-scientific firsthand experience. My point is that I am observing a new normal of several degrees' warmer temperatures in the 25,000' to 39,000' altitude range.

And then there are the superstorms. On our radar, we avoid red: red is bad, red is dense precipitation associated with thunderstorms. In my early years, especially during summer in the mid-west United States, we saw a lot of red. Lately, it's big patches of purple. This is a manifestation of what was formerly red, combining to form huge purple amoeba-type images on the airplane's radar screen, indicating super cells. I happened to be flying over Joplin, Missouri when it was devastated in 2011. I commented to my co-pilot "Jeez, I've never seen that before" such a big image on the radar screen, "Me, either," he said. After we landed in Dallas, Texas about an hour later, we saw the devastation on the news. Not long after that, it I was having dinner in a Birmingham, Alabama hotel restaurant. I commented to the wait staff that service was kind of slow. Breaking out in tears, the waitress said half of the hotel staff didn't come to work today because they died in last night's storm that destroyed suburban Birmingham. Many of the hotel staff lived in an area that was particularly vulnerable and was hit especially hard by the extreme weather. I felt very bad for complaining about the service. Then, on a flight from St. Louis to Dallas not long after the Birmingham experience, I noted a line of thunderstorms developing between our airplane and the St. Louis airport. It was bigger and more threatening than normal, even for the season. I got a message from the airline's dispatch department, first ever in my multi-decade career, to slow down because the St. Louis control tower had been evacuated due to tornados. I figured it would be a false alarm and the tower would be re-inhabited by the time we got there. A few minutes later, another first: I was told by the airline to return to Dallas. The St. Louis control tower was severely damaged, and the terminal was partially destroyed by tornados. Over the airplane's PA system, I told the passengers we were returning to Dallas for weather reasons in St. Louis. The passengers revolted, accusing the airline of some underhanded scheduling mischief. I relented and told the passengers the whole story of the devastation. The passengers calmed down and became concerned for their families. The whole experience was surreal for all of us.

In 2005, on an expedition to Antarctica in a vintage Russian power boat converted to an eco-tourism vessel, we had crossed Drake Passage, the body of water separating southern Chile's Tierra Del Fuego from Antarctica's Weddell peninsula, when we started encountering icebergs, some the size of office buildings.  As we approached landfall in Antarctica, we started to observe ice formations on land giving way and falling into the sea. These pieces of ice falling into the sea varied in size between small houses and apartment buildings.

Five years later, in the fall of 2010, as part of a sailboat crew transiting the North Atlantic between Iceland and Greenland, we were advised by authorities that an iceberg half the size of New York's Manhattan Island had broken free of the icepack and could be a hazard to navigation; no kidding! The largest Arctic iceberg in nearly half a century had split into two pieces after crashing into a rocky island west of Greenland. The iceberg — actually a massive floating ice island — broke off Greenland's Petermann Glacier. It was the largest such iceberg since 1962, with a surface area of 100 square miles and a thickness of about half the height of the Empire State Building.

When our sailboat finally made it to Greenland, we ran into a team of academics studying the rate of ice melt of the Greenland ice pack. I chatted with a researcher from the University of Copenhagen. I asked what he had learned from his research. He looked at me with a serious yet defensive look. "Listen," he said, "I don't want to debate about what's causing climate change, but I can tell you this. I've been studying these ice floes and core samples for decades and I can tell you, there has been a phenomenal increase in melting from 2003 to present, an unprecedented amount." (For further information see Dansgaard, W. (2005). Frozen annals: Greenland Ice Cap Research. Copenhagen: Niels Bohr Institute.)

Fast forward a year later to Puerto Williams, Chile in January of 2011. I'm in a bar, the *Micalvi*, a partially-sunken vessel that acts as a de facto yacht club for the Chilean military officers, transient "yachties", and a few scientific folks. I ran into Charlie Porter, a real character. Porter was an American, a Director of the Patagonia Research Foundation, adventurer and Captain of the Ocean Tramp. Charlie had many years' experience traveling and conducting climate science fieldwork along the southern coast of Chile and on the island of South Georgia. He had captained

**Antarctica: Iceberg falling**

Shutterstock

a number of expeditions to the glaciers and islands of Patagonia, including "Chronology of Late Holocene Moraines" (2006) and "Ice Cores from Patagonia" (2005), both in association with the University of Maine; "Isla de los Estados" (2005) with Lund University; and the "Scotia Centenary Antarctic Expedition" (2002) with the Royal Scottish Geographical Society. Charlie clearly had a deep academic background in terms of experience studying and actually investigating climate change through core sampling. I asked Charlie the $64,000 question: What's your opinion on climate change? He said, "Well, the amount of melting since 2003 is unprecedented." I ask, "What's causing it?" He replied, "Probably too much C02 in the atmosphere." I'm concerned: that's the same statement the Greenland scientists made, the part about unprecedented melting. I ask Charlie about the velocity of melting. "It's off the charts" he said, "The melting has never been this fast before, we have not seen this rate and velocity of melting for millions of years, based on core samples." "So, this is not a normal cycle?" I asked. "No, it's not," Charlie said.

**Checking weather enroute onboard sailboat**     Michael Day

# PART 1:

## BACKGROUND

# Chapter 1

# HISTORY OF HUMAN ADAPTATION
# TO CLIMATE CHANGE

The climate has been changing. It's normal. Not changing is abnormal. Climate change plays a strong role in the narrative of the Old Testament, for example. "So Noah, with his sons, his wife, and his sons' wives, went into the ark because of the waters of the flood..." (Genesis 7:7) Whether you believe the Noah story to be real or a myth, it's clear that climate has been changing — and civilizations have been adapting to climate change — for a long time. I view Noah's early adventure as a cultural metaphor for dealing with climate change, an early example of adaptation.

Throughout the history of our world, the climate has changed, and big climate change is the norm, not the exception. Recently, we've enjoyed an unprecedented time of relative climate stability, centuries long. Most scientists believe the dinosaurs were wiped out quickly by an event that spewed debris in the atmosphere causing a greenhouse effect. While scientists differ on whether the cause was a meteor impacting the planet or a volcanic eruption, the result was the same, a lot of debris in the atmosphere. Whether debris is caused by a meteor, a volcano, or hydrocarbons from industrialization, the effects are similar.

So what is "climate"? Climate is a long-term weather pattern, the weather conditions prevailing in an area in general or over a long period, such as a location known for its "cold, wet climate." A particularly cold winter is not climate change. Rather, climate change is generally considered 30 years of colder or warmer winters in a specific area. What happened to our ancestors and how did they adapt? Noah theoretically took to the sea, people from the Fertile Crescent left the area and moved to Europe and elsewhere. What will we do if global warming continues and sea level continues to rise? We will adapt.

The term "Fertile Crescent" was first coined in 1916 by James Breasted, an American archeologist and Egyptologist. The term refers to an ancient area of fertile soil and important rivers stretching in an arc from the Nile to the Tigris and Euphrates Rivers. It covers Israel, Lebanon, Jordan, Syria, and Iraq. The Mediterranean lies on the outside edge of the arc. To the south of the arc is the Arabian Desert. To the east, the Fertile Crescent extends to the Persian Gulf. Geologically, this corresponds with where Iranian, African, and Arabian tectonic plates meet (Gill, 2020).

Traditionally associated with the Christian, Jewish, and Muslim faiths, the Fertile Crescent has been said to be the Earthly location of the Garden of Eden. It was the cradle of civilization and is considered the birthplace of agriculture, urbanization, writing, science, history, and organized religion circa 10,000 BCE (Mark, 2018). At its peak, the Fertile Crescent was the agricultural hub of the Middle East. But the Fertile Crescent is not so fertile anymore. Having spent some time there, I have seen that this area is strongly challenged by a dry and arid climate. The rich Mesopotamian marshlands known for centuries have almost completely disappeared, with only 10 percent of the important ecosystem still remaining, according to a study based on satellite images of the region (NatGeo 2010).

Another example of climate change in the neighborhood of the Fertile Crescent is Syria. There is compelling evidence that the Syrian War was exacerbated by the poverty that resulted from the severe drought of 2007-2010, which put those in the agriculture business into despair. The war displaced over 11 million people, half of the 22 million of the prewar population. Most of the refugees were displaced to Turkey, Lebanon, Jordan, Iraq, and Egypt (Stamm & Harness, 2016). This migration put pressure on the resources on these countries, especially Turkey. While the migration of refugees into Turkey may ultimately prove beneficial to Turkey, the impact of more people being so quickly added to an existing population will create stress to an already stressed economy. This is a case of adaptation in progress.

We see another example of adaption (in this case, adapting by leaving an area no longer habitable due to natural disaster) in Akrotiri, on the Greek island of Thera (currently referred to as Santorini). The island, which some think was Plato's Atlantis, disappeared from the Greek landscape due to the Theran volcanic eruption of 1627 BC. Many of the Akrotirians actually escaped, likely to Crete and elsewhere. Today there is a very interesting and well-preserved archeological site there near the town of Santorini.

Volcanic eruptions are smaller examples of global or semi-global climate change. What's happening today with global carbon output into the atmosphere is similar to the concept of volcanic debris, only more intense. Volcanoes cause greenhouse gases, which make the climate warmer. Hawaii's Kilauea volcano is a recent example of debris spewing into the atmosphere. The particulates in the air of the Big Island of Hawaii had created "vog," a fog created from moisture in

**Akrotiri, Archeological Site**                    Photo by Michael Day

the air grabbing onto the volcanic debris. This fog affects air travel, makes people with respiratory issues sick, and negatively impacts tourism in Hawaii. This is a very contemporary example of environmental cause and effect. Hawaiians have been adapting to volcanoes for a long time. They had traditionally rebuilt quickly with relatively impermanent housing such as huts. More recently, houses are more contemporary ranch style homes which get destroyed and rebuilt on the same land near the volcano, over and over. This is another form of adaptation, in this case not abandoning the homeland.

We can even see this on a planetary scale. It would be logical that Mercury would be the hottest planet in the solar system since its closest to the sun. But Venus, while farther from the sun than Mercury, is hotter because of greenhouse gases. These gases are composed of carbon dioxide ($CO_2$) and sulfur dioxide ($SO_2$). The average temperature on Mercury is 332 °F (167 °C) while Venus is 864 °F (462 °C) (Redd, 2012). Earth's average temperature is roughly 59 °F (15 °C). Venus is hotter, a lot hotter, because of greenhouse gases. If Earth's atmosphere continues to be polluted by particulates causing more and more greenhouse effects, we could ultimately suffer the outcome of Venus.

Climate modeling is based on global atmospheric patterns similar to those that occurred in the past. Our most precise knowledge comes from Earth's paleoclimate, its ancient climate, and how it responded to past changes, including atmospheric composition. Data found in tree rings, ice cores, and lake sediment yield our best assessment of what might happen to the climate in the future. Our second essential source of information is global observations today, especially satellite observations, which reveal how the climate system is responding to rapid human-made changes of atmospheric composition, especially atmospheric $CO_2$. Models help us interpret past and present climate changes, and, insofar as they succeed in simulating past changes, they provide a tool to help evaluate the impacts of alternative policies that affect climate.

**Venus**

By vsop | Shutterstock

Ultimately, the global temperature change in response to energy the Earth receives from the Sun versus what's radiated back into space, known as "climate forcing," is the factor that drives climate change. Climate forcing changes the Earth's energy balance, either through a change of the sun's human-made change of atmospheric $CO_2$. Some is doubling the atmospheric $CO_2$ levels, because that is the level that humans could impose this century if fossil fuel use continues without restriction.

Humans lived in a rather different world during the last ice age, which peaked 20,000 years ago. An ice sheet covered Canada and parts of the United States, including Seattle, Minneapolis and New York City. More than a mile thick on average, this ice sheet would have towered over today's tallest buildings. Over time, our planet went through a series of climate shifts, from glacial to interglacial periods, during which atmospheric $CO_2$ levels decreased and increased. These natural freeze and melt processes were relatively slow, and the effects they had on the planet's atmosphere pale in comparison to the effects of human-caused $CO_2$ (Hansen & Sato, 2011).

As examples of non-human impact, let's look at some other famous volcanic eruptions. Volcanic eruptions create and destroy landscapes, fascinate and terrify observers, and affect our climate.

## Mount Vesuvius, Italy—AD 79

Mount Vesuvius is one of the world's most dangerous volcanoes. Mount Vesuvius is in Naples, Italy, and remains one of the nastiest volcanoes in the world, exacerbated by the fact that a population of 3,000,000 live in close proximity. The deadly power of Vesuvius was dramatically demonstrated in AD 79, when a monumental eruption wiped out the Roman cities of Herculaneum and Pompeii.

Vesuvius had been showing signs of activity for several years before the great eruption, the most dramatic of which was a powerful and destructive earthquake 17 years previously. Earthquakes and volcanoes commonly occur in tandem, a phenomenon which would not have been lost on the Roman population. It is now known that one of the main causes of both earthquakes and volcanic eruptions is the movement of the Earth's tectonic plates. These plates move both away from each other and towards each other, causing pressure to build up inside the volcano, which in turn leads to the movement of the magma within it.

It is the release of this pressure that ultimately causes earthquakes and eruptions. Although the connection between the two would have been observed by the Romans over time, the perpetual nature of volcanic activity in this region meant that the warnings were largely unheeded by the population.

On August 24, AD 79, Vesuvius erupted, spewing a column of ash and pumice high into the atmosphere around the volcano. The column rose more than 15 miles and was carried toward the cities of Pompeii and Herculaneum by strong easterly winds. Within hours, the cities were buried in yards of ash, and the volcanic cloud that accompanied the eruption blocked out the sun completely, leaving the area in complete darkness. The scene must have appeared almost apocalyptic to the cities' terrified inhabitants—but it was to get worse.

Between midnight and daybreak on August 25th, a series of six devastating episodes of matter and gas spewing into the atmosphere, described as a glowing avalanche of hot ash known as pyroclastic surging, wreaked devastation beyond comprehension. The first powered down the volcano at speeds of up to 62 miles per hour, wiping out everything in its path. The second surge reached Herculaneum, burying it completely, and the fourth surge reached as far as Pompeii, laying waste to the city before the population had time to flee. Recent excavations have found hundreds of bodies preserved in the volcanic material, with fear and desperation twisted on their faces for eternity. Nobody knows the exact number of people killed by that eruption of Vesuvius, but it is believed to be in the thousands (Sigurdsson, 1999).

People eventually returned to the areas in and around Naples near Pompeii and Herculaneum. The population of Italy overall has thrived over the years since these disasters. The survivors, those far enough away from the volcano's effects and their descendants, adapted to the new post eruption environment and thrived.

All challenges are not driven directly by physical impacts such as volcanos and long-term climate change. Social changes can be equally difficult, or even more devastating than physical ones. New challenges are continually presenting themselves to modern day Europe, including Italy, such as an aging population, chaos within the European Union, and economic pressure from the recent onslaught of refugees. Society has adapted by sharing the resources and responsibility between members of the EU, and by continually restructuring debt. Italy's debt will

**Eruption victim of Mt. Vesuvius in Pompeii**  © Floriano Rescigno/Dreamstime

likely be restructured again in the near future as the cumulative impact of Italy's economic problems and the COVID-19 pandemic take their toll (Reuters Staff, U.S. Markets, June 8, 2020). While controversial and painful, these challenges are so far being met through continual adaptation, keeping Italy a global economic player.

The 1982 eruption of Mount St. Helens caused economic devastation in the region. In 1978, volcanologists Dwight Crandell and Donald Mullineaux warned that Mount St. Helens would erupt before the end of the century. On May 18, 1980, Crandell and Mullineaux's predictions came true with profound consequences (Francis, 1993). The disaster began at 8:32 that morning with a large earthquake measuring 5.1 on the Richter scale and triggering a series of explosions that increased in frequency and intensity. The eruption reached its peak with a violent explosion around 3:50 pm, causing the largest debris avalanche ever recorded to power 14 miles west. On top of this, a large cloud of ash rose up to 15 miles into the atmosphere. The combined results of the seismic activity throughout the day devastated the landscape (Brantley and Myers, 2005).

The environmental cost was one thing, but the human cost was the main trag-edy of Mount St. Helens. Moments after the main avalanche, an explosive blast ripped through the area, destroying everything in its path and killing 57 people. The eruption caused more than $1 billion worth of damage, primarily to the lum-ber and agricultural industries. Forests, roads, bridges, recreational sites, houses, trails, railways and wildlife habitats were damaged or obliterated. Tourism to the region fell and unemployment rose in the immediate aftermath of the disaster. The economic consequences for a region which supplies lumber and agricultural resources not only to the Pacific Northwest but also to the whole country were huge (Tilling, Topinka, Swanson, 2005).

But nature and society adapted. Scientists are stunned at the area's recovery, noting 150 species that have returned to the "lifeless area." (NatGeo) The econ-omy of Washington has grown exponentially since 1980, mostly due to the tech

**Petrified Tree And The Devastated Landscape**
**Of Mt St. Helens From The 1980 Eruption**
© Roodboy | Dreamstime

and real estate booms. Following $1 billion in damage caused by the natural disaster, billions of dollars were created by real estate growth in the decades to come. Not only did the state of Washington recover from the Mt. St Helen's eruption, but it thrived through unprecedented economic growth that overtook its pre-eruption economy.

The almost un-pronounceable name of Eyjafjallajökull was to become one of the most well-known volcanoes in the world when the enormous ash cloud it produced in Iceland wreaked unprecedented travel chaos throughout Europe and the world, causing billions of Euros of damage to the fragile European economy. Between March and May 2010, the volcano spewed ash into the atmosphere, forming a thick, black cloud. Strong winds carried the cloud as far as mainland Europe, and for several weeks, flights were canceled or disrupted, major airports ground to a halt, international sporting events were cancelled, and the struggling

**Eyjafjallajokull volcano eruption in Iceland, 2010**    By J. Helgason | Shutterstock

economy suffered a further blow. Around 100,000 flights whose paths would normally cross over the area were canceled, affecting more than 10 million passengers. The European Union Transport Commission estimated that the ash cloud may have cost European businesses up to €2.5billion (Gabbatt, 2010).

I was part of a crew sailing on an 80-foot sailing vessel from Iceland to Greenland in 2010. We knew that we might be in trouble from the erupting volcano. European travel was being decimated by the volcano. After closely checking the weather, wind directions, and the expected volcanic ash track, we concluded that we could safely make the Atlantic crossing to Greenland even though during the same period airliners could not land at London's Heathrow airport. We were able to sail from New York to Reykjavik, from Iceland to Greenland, and up the West Coast of Greenland without incident, in spite of Eyjafjallajökull.

All of these volcanic events had impacted the people that lived in the vicinity of the eruption, and farther away. Some of the people were wiped out, some managed to stay and adapt to a new normal, while others emigrated. These options have historically been the main course of behavior following a natural disaster. While significant global climate change may make emigration less possible or even impossible in the future, population shifts in the relative short term are a likely consequence.

Humans cannot control volcanoes, but there are certain causes of climate change that we do have some influence over. According to a study performed by Columbia University on the long-term health effects of the bombing of Hiroshima and Nagasaki in World War 2, between 90,000 and 166,000 people died in Hiroshima, and 60,000 to 80,000 people died in Nagasaki within the first few months after the bombing. Seventy years later, most of those that survived the bombings have died. One of the most immediate concerns after the attacks regarding the future of both Hiroshima and Nagasaki was what health effects the radiation would have on the children of survivors conceived after the bombings. According to the Columbia University study: "...So far, no radiation-related excess of disease has been seen in the children of survivors, though more time is needed to be able to know for certain. In general, though, the healthfulness of the new generations in Hiroshima and Nagasaki provide confidence that, like the oleander flower, the cities will continue to rise from their past destruction." (Listwa, 2012)

There is no doubt that these bombings were horrific and unconscionable, even during a global war. The big fear after a nuclear explosion are fission products. Most of these were dispersed in the atmosphere or blown away by the wind. Though some did fall onto the city as black rain, the level of radioactivity today is so low it can be barely distinguished from the trace amounts present throughout the world as a result of atmospheric tests in the 1950s and 1960s. The other form of radiation is neutron activation. Neutrons can cause non-radioactive materials to become radioactive when caught by atomic nuclei. However, since the bombs were detonated so far above the ground, there was very little contamination—especially in contrast to nuclear test sites such as those in Nevada. In fact, nearly all the induced radioactivity decayed within a few days of the explosions.

Today, the liveliness of the cities of Hiroshima and Nagasaki serves as a reminder not only of the human ability to regenerate, but also of the extent to which fear and misinformation can lead to incorrect expectations. After the bombings of Hiroshima and Nagasaki, many thought that any city targeted by an atomic weapon would become a nuclear wasteland. While the immediate aftermath of the atomic bombings was horrendous and nightmarish, with in-numerable casualties, the populations of Hiroshima and Nagasaki did not allow their cities to become the sort of wasteland that some thought was inevitable. This experience can serve as a lesson in the present when much of the public and even some governments have reacted radically to the more recent accident in Fukushima: in the midst of tragedy, there remains hope for the future.

These natural and man-made disasters, while devastating, are all smaller scale, temporary samples of micro-climate change. Environments were decimated for different periods of time and the affected civilizations adapted in various physical and cultural ways to survive and ultimately thrive. However, devastation on a global scale presents different problems: there is nowhere to escape to and not enough alternative climates to help absorb the impacts of other more distressed areas. On a global scale, climate change has a larger, more powerful, and interconnected impact.

# Chapter 2

# THE SCIENCE OF
# A CHANGING CLIMATE

Of course, this is not mankind's first rodeo with climate change. Over the past 5 to 7 million years, Homo sapiens have experienced, survived, and learned from many climatic shifts. But the current one is unique. Why is that?

In January 2019, the Harvard Business Review published The Story of Sustainability in 2018: "We Have About 12 Years Left" by Andrew Winston. Winston is the go-to expert on how business can profit from responding to a warming planet. In this publication, Winston concluded that 8 realities exist:

1.  The world's scientists sound a final alarm on climate.

2.  Antarctica's melting 3 times faster than a decade ago. Greenland is losing ice quickly, causing potentially unprecedented sea level increases which could reach 200-plus feet.

3.  Entire towns are being wiped off the map by extreme weather. They range from Paradise, California to Mexico Beach, Florida.

4.  Coral is dying. Insects are disappearing. The fate of major ecosystems looks catastrophic.

5.  More investors are viewing climate and sustainability as core value issues.

6.  The clean technology explosion continues and accelerates.

7.  China rejects the world's trash.

8.  Meatless options grow plentiful.

I have piloted commercial jets for major airlines for over three decades and I have noticed that the temperatures aloft have become warmer. When I started in the 1980s, the temperature deviation varied from -7 °C to +7 °C. Essentially, temperature deviation represents the difference between what the temperature is at a specific altitude versus what it should be, based on normal or standard temperature. At sea level, the normal temperature is defined as 15 °C or 59 °F. As we ascend, the temperature decreases at about 2 °C per 1,000 feet, which is the "adiabatic lapse rate." It's important to know what the temperature is at a certain altitude, compared to sea level, in order to accurately assess the aircraft's ability to perform at higher altitudes. Let's put it this way: the engines and aerodynamic components such as the wings care a lot about temperature. As I have pointed out earlier in this book, during the past decade I have experienced a lot fewer minus numbers and lots more plus ones. They have often been in the +10 °C to +12 °C above normal range.

What had been making headlines in the media at the time was the 2 °C sea level increase threshold. What I noticed was that the increase was much higher at higher altitudes. That was stunning. It also was the tipping point in my determination to uncover how investing would be disrupted by those developments.

This was all a shock. In 1980, when I was studying climate, I assumed that globally we were in a reasonable pattern of climate variation. There were warnings that if carbon in the atmosphere continued to increase, that would cause warming. But at the time, the warning didn't seem dire. The red flags began to appear with the big spikes, especially at higher altitudes, in the late 1980s. Those increases continue to accelerate.

This is the era of so-called "fake news." Therefore, we could be skeptical. We could wonder if climate change is real. Unfortunately for mankind, researchers have documented that the current climate situation and how the future could play out are not simple standard cyclical activity. Back in school, we studied the scientific method. We recall how rigorous it is and how many controls are used. Both in the U.S. and globally, independent scientists are struggling to understand the dynamics of climate change through a variety of methodologies. Among the most compelling investigations are those by scientists such as climate expert Allegra LeGrande of the NASA Goddard Institute for Space Studies and physical scientist Kimberly Casey of the U.S. Geological Survey and Cryospheric Science Lab at NASA's Goddard Space Flight Center. Their focus has included the study of ice cores. Ice cores are analogous to tree rings in that they tell us a lot about the history, current health, and future fate of an environment.

To analyze ice cores, scientists drill into glaciers and ice sheets near the earth's North and South Poles. Those were formed from years and years of accumulating snowfall. Annually, the weight of the new snow compresses the previous layers of snow. What results are ice sheets that can be several miles thick. Each sheet of ice tells a story about what the earth was like when a particular layer of snow fell; for example, the temperature of the air during a specific era, global events such as volcanoes, and atmospheric amounts of greenhouse gases such as $CO_2$ and methane.

But the benefit of the data isn't only historic. Based on temperature information locked in the ice core, it's possible to predict Earth's future climate. As a result of her studies so far, LeGrande has become certain of one thing. She put it this way: "The climate of the next century will be well beyond the range of the climate that we have observed for the past 160 years." (Stoller-Conrad, 2017). I put it this way: "The past and the present won't be the future."

Not all gases are created equal. Some have greater negative impacts on ecological balance, physical survival, and economic well-being than others. $CO_2$, methane, and water vapor trap heat more than other gases. An increase in their presence means more absorption of heat, and that isn't good. Paleoclimate research has shown that $CO_2$ has a strong correlation to temperature: shifts in temperature and $CO_2$ levels are closely linked. Scientists have also analyzed the results of these major climate shifts, which benefit some organisms and are fatal to others. Here's a bit of history. $CO_2$ levels per million (ppm) had remained relatively consistent over the past 800,000 years, varying between 150 ppm and 250 ppm. Humans have been living in a climate conducive for supporting huge biodiversity in organisms for the past 10,000 years. $CO_2$ levels and temperatures have been relatively constant as mankind moved into an interglacial period

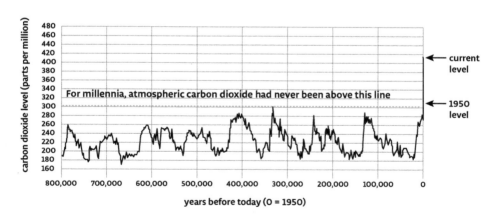

Chart courtesy of NASA's Goddard Institute for Space Studies (GISS)

approximately 12,000 years ago, during which (10,000 years ago) the birth of civilization coincided with the birth of agriculture. Agriculture allowed mankind to settle down and create institutions, ranging from art to higher education. But $CO_2$ levels are no longer constant. The Industrial Revolution rapidly changed this consistency to the point where $CO_2$ levels have now reached 400 ppm, a level not seen for over 1 million years. If history proves itself telling as it usually does, temperature will follow.

**What does this increase in $CO_2$ mean? Let's take a look at the greenhouse effect.**

Certain gases reflect shortwave and longwave radiation and certain others absorb them — some better than others. Visible light and ultraviolet light are commonly called shortwave radiation. Infrared radiation is referred to as long-wave radiation. The sun radiates energy mainly in the form of visible light, with small amounts of ultraviolet and infrared radiation. For this reason, solar radiation is usually considered shortwave radiation. The Earth produces virtually no visible light, or ultraviolet radiation; almost all of the natural radiation created by the Earth is infrared, or heat. Thus, we refer to terrestrial radiation as longwave radiation. It is this longwave radiation which exacerbates the greenhouse effect.

Both $CO_2$ and water vapor absorb the longwave radiation emitted by earth much more efficiently than any other gas. $CO_2$ and water vapor are linked because as $CO_2$ levels increase, temperatures increase, which in turn increases the amount of precipitation, thereby increasing water vapor. While water vapor can produce a cooling effect in large quantities due to reflection, it also traps the longwave radiation coming from Earth. $CO_2$ and water vapor are the main way that earth maintains climatic homeostasis.

Climatic homeostasis or equilibrium is essentially based on the carbon cycle. Carbon is sequestered or segregated and released by natural processes. The sources of sequestration? Among them are sedimentation, photosynthesis, silicate weathering, and reforestation. The reasons for the release? They include volcanism, use of fossil fuels, methane release, deforestation, wildfires, and respiration. Yes, these processes can vary in time from hundreds to millions of years. Unfortunately, the sources of release have increased significantly over the past century. That has been shown to be from the uncovering and burning of fossil fuels. Using fossil fuels is not totally negative but using fossil fuels at the current rate is alarming. The carbon cycle takes thousands to millions of years to achieve balance and humans are altering that process significantly.

Paleoclimate research has revealed unanticipated insight into how $CO_2$ and temperature correlate to certain climates, based on solar intensity, geographical positioning, and ocean circulation. According to the Niels Bohr Institute at the University of Copenhagen, future global warming will depend not only on the amount of emissions from human-made greenhouse gases; it will also be linked to the sensitivity of the climate system and response to feedback mechanisms. By reconstructing past global warming trends and the carbon cycle on earth 56 million years ago, researchers from the Niels Bohr Institute, among others, have used computer modeling to estimate the potential for future global warming, which could be greater than previously thought.

We've discussed the atmosphere's role in the greenhouse effect: what about our oceans? There is much more to climate change than what is occurring in the oceans, but oceans grab attention because they consume about 70% of atmospheric $CO_2$. As atmospheric $CO_2$ continually increases, the oceans become saturated and unable to continue their filtering. What then happens is oceans and atmosphere are at such similar concentrations of $CO_2$ that transference cannot happen. Consequently, $CO_2$ is transformed into carbonic acid in the oceans. Carbonic acid breaks down calcareous organisms such as coral and has a direct negative impact on the oceanic food web as a whole.

Now, to move to back to the climate as a whole, it's no coincidence that, during the past few years, the U.S. has endured 25 climate-and weather-related disasters, including firestorms, droughts, heatwaves, hailstorms, hurricanes, and tornadoes. According to the National Oceanic and Atmospheric Administration (NOAA), these disasters claimed 1,141 lives and generated $175 billion in damages. In response to these climate-change catastrophes, NOAA has established 5 goals:

1.  Reduce vulnerability to extreme climate and weather events,

2.  Prepare for droughts and long-term water shortages,

3.  Protect and preserve coastal resources,

4.  Identify and manage marine resources, and

5.  Plan ways to adapt to climate change while also mitigating its impact.

In the Paris Agreement on climate change in 2015, signatory nations pledge to keep a global temperature rise in the 21st century below 2 °C. The strategies to accomplish this include global cooperation, a new technology framework, raising $100 billion in funding, and enabling developing economies to get into the act.

There seems to be a global paradigm shift in individual consciousness about how to live a life on now-distressed planet earth. "Crazy weather" is a term being heard increasingly around the world. That is just a small part of the impacts of climate change. Here are others:

**Wildlife.** The thinning polar ice pack, melting glaciers, rising sea levels, decreased water salinity, and increased methane release are creating new kinds of endangered species. They include the Caspian seal, fin whale, and polar bears. As they vanish, their ecosystems become even more unbalanced, potentially leading to more species becoming endangered.

**Agriculture.** Rising temperatures will reconfigure what can be grown where. Starvation could come to certain regions. Political tensions could escalate, both internally and among nations. Will society return to a nomadic pattern in its search for food? The institutions it has put together could collapse in another dark age.

**Breaking Off and Melting of Ice Sheets.** Ice melt creates a circular process, as melt speeds up the glacial flow into the ocean which, in turn, accelerates the melting. In addition, there will be wide-scale release of pollution that had been trapped in the melting ice.

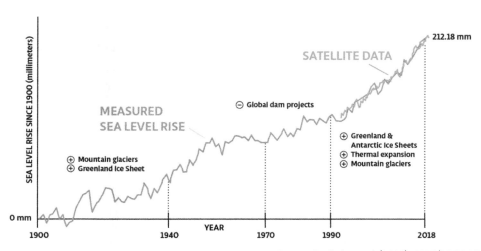

Source: Frederise et al. (2020); GSFC/PO.DAAC

**Sea Level.** Conservative projections for the next 80 to 100 years show a rise of 2.67 feet.

This infographic shows the rise in sea levels since 1900. Pre-1940, glaciers and Greenland meltwater dominated the rise; dam projects slowed the rise in the 1970s. Now, ice sheet and glacier melt, plus thermal expansion, dominate the rise. Tide-gauge data shown in blue and satellite data in orange.

**Thermal Expansion.** As temperature increases, molecules begin vibrating faster and increase their distances from one another. As ocean temperatures increase, the volume of the ocean will increase through heat transfer. Damage has already been done: according to a 2017 study by researchers at MIT and Simon Fraser University, even if the world were to stop emitting $CO_2$ starting in 2050, up to 50% of the gas would remain in the atmosphere for more than 750 years afterward (Jennifer Chu | MIT News Office, 2017). That means that post-$CO_2$ emissions, the rise in sea level would continue to increase. Some put that at twice the level of 2050 estimates for 100 years, and 4 times that value for another 500 years. According to NOAA updated 11/05/2020 sea level is rising at about 1/8 inch per year.

Why? According to Susan Solomon, the Ellen Swallow Richards Professor of Atmospheric Chemistry and Climate Science at MIT, the acceleration is due to "ocean inertia". As the world warms due to greenhouse gases — $CO_2$ included — waters heat up and expand in volume. Lowering sea levels by removing the extra ocean heat caused by even short-lived gases is an extremely slow process. "As the heat goes into the ocean, it goes deeper and deeper, giving you continued thermal expansion," Solomon explains. "Then it has to get transferred back to the atmosphere and emitted back into space to cool off, and that's a very slow process of hundreds of years" (quoted in Chu's article). The Natural Sciences and Engineering Research Council of Canada and NASA supported those findings.

Everything changes. Right now, and for several centuries to come, mankind will continue to adapt to climate change and struggle to mitigate its impacts. International, national, and local policies will play a significant part in that. Meanwhile, we can spot and harness fresh opportunities on how to build, preserve, and grow wealth. It was management guru Peter Drucker who observed that in turbulence amazing opportunity can emerge.

# PART 2:

## SECTORS

**Chapter 3**

# FOOD'S ECONOMIC IMPACT
# AND THE FUTURE OF EATING

I am a food explorer. There is nothing I won't try — nothing! At the Explorer's Club annual dinner in New York there is an exotic food table that is not for the faint of heart. Past dinners have included such fare as tarantulas, fried cockroaches, deep fried earth worms, steamed jellyfish, durian (allegedly the smelliest plant in the world), camel burgers, spleens, goat genitals, and various eyeballs. Not everything is available every year, and now only sustainable foods are available, nothing endangered is served. When my wife and I were there a few years ago, I tried at least one of everything that was available: my wife did not. I liked some things better than others. The worms were a little too mushy for my taste.

**Durian**

People need food, people love food. There are food channels, food blogs, mainstream movies about food and cooking such as Chef; The Hundred Foot Journey; Jiro Dreams of Sushi; The Lunch Box; Eat Drink, Man Woman; Julie and Julia; and Chocolat. There is the Food Network, The Great British Baking Show, and the movie Chef has spawned a TV show called Chef Show. We have clearly become a food culture. In fact, I would argue that food is equal to water in terms of sheer importance to humanity.

Globally, we produce more food than ever. Just prior to the early 1800s the world's population was in the millions; today it's almost 8 billion. Early humans scraped for food, they were often hungry and died of starvation. Wars were fought over food: there were the "Ottoman-Venetian Wine Wars" of 1570-1573, the "The Pig War" of 1859 between the British and Americans in Canada, and the "Pastry War" of 1838 between France and Mexico. People take their food seriously.

The food industry is huge. Agriculture, food, and related industries contributed $1.109 trillion to U.S. gross domestic product (GDP) in 2019, a 5.2 % share. The output of America's farms contributed $136.1 billion of this sum—about 1 percent of GDP (USDA.gov). The food industry is evolving into more local production, such as the farm-to-table movement. Organic foods and plant-based foods are becoming mainstream. Amazon bought the Whole Foods grocery chain, institutionalizing a paradigm shift from old-school grocery delivery to a new normal of fresher produce and a more environmentally aware and health-conscious supply chain of food delivery.

There are places in the world that suffer from famine and food shortages. However, the developed world is suffering from an overabundance of food and obesity, not food scarcity. When resources are abundant, we can afford to over-use or waste them. When resources are limited, we have to be smarter and more careful. The basic economics to conserve resources comes into play. Globally, we waste about one-third of all the food that is produced. Can this situation last, an overabundance of food? Yes and no.

The production of food requires water, chemicals, and energy. If water becomes scarce, food will also be scarce. Chemical fertilizers and pesticides are standard components of modern farming. Energy is a problem in terms of electricity and fossil fuels: we've discussed the harm fossil fuels can wreak on our climate. We've arguably reached a tipping point of negatively impacting the climate by the way we produce food, especially meat, and even worse from larger animals. In terms of air pollution, toxic methane greenhouse gas is produced by the meat and dairy industry, primarily cattle passing gas, with a warming potential nearly 30 times greater than $CO_2$.

The difference between the water required to create one pound of meat versus a pound of vegetables is profound. It takes approximately 15,415 liters of water to produce 1 kilogram of bovine meat, while one kilogram of vegetables requires only 322 liters of water. Bovine meat takes over 47 times more water than vegetables (Mekonnen and Hoekstra 2010). The obvious consideration is for food animals to eat vegetables: if the ranch animals eat plant-based feed, the water needed to grow that feed combined with the water used for animal drinking water, to clean up after the animals, and the water associated with butchering creates an increased aggregate water usage (watercalculator.org).

While arguably more sustainable than meat production since the animal does not have to be killed to produce milk, the dairy industry is still in trouble. Dairy alternatives such as soy, oat, and almond milk have helped to put 1600 dairy farms, in the state of New York alone, out of business between 2006 and 2016. Annually, milk production has slightly increased nationally but consumption of milk has slowed (American Farm Bureau Federation, fb.org 2020). While dairy production is slightly increasing, the writing is on the wall for an industry downturn based on the alternatives for dairy that appear to be healthier options.

Land use is another big factor affecting food production and availability. Amazon River land has recently been repurposed from rainforests to ranching. While this action has created opportunity to raise more livestock, repurposing rainforest has also exacerbated global warming and reduced the oxygen in the atmosphere (Malhi et al., 2008). The carbon footprint of beef cattle raised on deforested land such as the Amazon rainforest is twelve times greater than from cattle raised on natural pasture (Nassos Stylianou, 2019).

The long-term answer may be more vegetable production in lieu of meat production combined with hydroponics. A subset of hydroculture, growing plants without soil, hydroponics is not new, but the way we use it is. In 600BCE the Hanging Gardens of Babylon were functioning through a pulley-system accessing water from the Euphrates river. This was an ancient form of hydroponics (Lagomarsino et al., 2019). Today, the roots of plants are suspended in water and sometimes secured by volcanic rock or gravel. The plants are fed nutrients from the water they are suspended in. They are fed with fertilizer, fowl and fish excrement, among other things. Tomatoes, peppers, cucumbers, lettuce and cannabis are commonly grown hydroponically. The process uses about 80% less water than traditional growing techniques and can allow food to be grown indoors in larger multilayer vertical structures, not unlike an office building. This process can

**Hydroponics vegetables growing in plastic pipes**     Boophuket; Shutterstock

bring food to difficult places where agriculture could not be traditionally grown because of harsh climates.

Bowery Farming is a working hydroponic farm in New Jersey. Bowery Farm can grow 365 days a year and uses 95% less water than the average agricultural farm. They grow without the use of pesticides or agricultural chemicals. According to Bowery, they grow 100 times more than the same square feet of a typical farm. In 100 years' time, maybe a large amount of food won't be grown in soil at all.

Dry farming, growing food without irrigation, is another new technique based on an old technology. The technique does not involve irrigating crops, relying instead on the crops' roots to go deep underground for water. Dry farming has been popular in Europe and is used with success in California today. According to a National Geographic article, John Williams from Frog's Leap Winery in Napa said (regarding the California drought) that "...if he didn't watch the news, he would not have known there was a drought." (Levaux, 2016)

Diversified farming is another concept in "new school/old school" farming. The idea is that smaller to mid-sized farmers can compete with large agribusiness farms by rotating crops and using their land for ranching as well. By being flexible with their planting and harvesting regimen, the farm can work around droughts and intermittent rain increases, or near flood conditions, by planting crops appropriate for those conditions. Some believe that the secret is in the soil so, by rotating crops, avoiding chemicals, and using natural compost, these smaller agricultural businesses can save money by not buying chemicals and at the same time growing organic crops with a higher profit margin. Organics is not just about ecology: it can be simply a good economic decision. Right-sizing the space for a given crop, spending less on chemicals, and getting a higher price based on the demand of organic products can actually give these smart and nimble farmers an edge over the large agribusiness models.

### Cultured Meat

Cultured meat, meat grown artificially at a cellular level, is a new technology to grow meat from fat or muscle cells. Beef or chicken, the meat is grown in a lab versus raising the whole animal for the ultimate purpose of providing food. The growing cells are put into a bioreactor to help make the cells grow. A bioreactor is a vessel that propagates biological reactions; it's a cell cooker. While this technology may not sound appetizing, it's effective and reduces the cost, by a lot. Future Meat's founder Yaakov Nahmais said the cost could come down from $800 per kilo to $10 per kilo (CNBC.com Amelia Lucas, Oct 10, 2019). Not only will cultured meat save money but it could save resources such as water, feed, and fertilizer.

**Protein: The myth**

So, if meat is so costly to produce, in terms of water, feed, and land, why are we still doing it? It's generally accepted that people need to eat a lot of protein and the protein comes from meat. That's not the whole story, though. Protein is essential. Proteins are broken down in the stomach into amino acids and are absorbed in the intestine. Many of us consciously eat high protein diets, but we don't need as much protein as we generally consume. We need about 50 grams of protein daily: that's about two palm-sized portions of meat, fish, tofu, nuts, or legumes. The average American eats over 100 grams of protein a day, twice what is required. Too much protein, over 2 grams per kg of body weight per day, can cause digestion problems, dehydration, and kidney problems. And protein doesn't have to come from meat or dairy products: there is a lot of protein in vegetables such as beans, corn, broccoli, avocado, asparagus, and potatoes, to list a few. Many of us can probably eat less protein and be healthier, with a bonus of saving money. Industry could benefit from a shift to a more plant-based diet. As resources become scarcer, the industries that make cheap and efficient foods could outpace less efficient technologies.

**The benefits of plant-based food**

From a personal standpoint, as a middle-aged male, I suffered from elevated blood pressure and an above normal "bad" cholesterol count. I got tired of fighting to keep my numbers down by reducing carbs, eating less, and watching my diet like a hawk, so I decided to embrace a primarily plant-based diet. I had done this once before in the early 2000s, but I was discouraged when I had trouble doing high calorie activities such as mountain climbing. Performing normal daily tasks was fine, but when I was training to climb Mt. Rainer in Washington State, I was running out of steam. The calories burned in an 8-hour day of climbing amounted to about 3600: added to the normal 2500 calories a day, I needed about 6100 calories a day. My plant-based intake was about 2500. I was calorie deficient. Rather than try to figure out how to boost my calories through my desired plant-based regimen, I grabbed what was handy such as chicken, eggs, etc. My mountain climbing exploits went well, but I gave up on my plant-based diet for a while.

I recently tried a plant-based diet again, this time with success. My blood pressure and cholesterol are way down, my weight is stable, I have more energy, and as a bonus, I have saved a fortune in expensive dinners. The difference this time was a combination of motivation and the availability of more mainstream plant-based food options than ever before. There are also many more plant-based competitive athletes at the top of their games in running, fighting, rowing, weightlifting, skiing, race car driving, and mountain climbing.

Dr. Dean Ornish, famous cardiologist and one-time doctor for President Bill Clinton, suggests plant-based diets as a way to turn heart disease around. One compelling example he discusses is a patient that was scheduled for a heart transplant. Dr. Ornish asked the patient's doctor if his patient could participate in Ornish's clinical trial of his plant-based regimen to see if a plant-based diet could help the guy. The patient's doctor asked Dr. Ornish how long the study was. Dr. Ornish said, "A year." The guy's doctor said, "...my patient probably won't last a year." The patient did participate in the study, and got significantly better after going on a plant-based diet. In fact, he didn't need the heart transplant and lived many years. But Dr. Ornish is not a zealot. He suggests that even strategies such as Meatless Mondays could go a long way to improve health and could destress the need of using scarce resources.

Food is a big deal. People often say that gold is a good financial hedge to a bad economy or a devastating crisis. I tend to disagree; I think food and water are the best hedge. Food is so important that at times people go nuts and hoard food when there is a perception of scarcity. We've seen this in the recent pandemic. As food becomes more difficult to produce, especially meat, things could become volatile. During the COVID-19 crisis, pork became difficult to produce because workers were transmitting COVID-19 to each other. The outcome was less pork available to the market, which caused higher prices and panicked consumers.

The food industry is tricky, there are a lot of competing factors. First, people need food to live. Food can taste good, be satisfying, and provide a center for social activities. Beside simple nutrition, food can have direct medical implications on people with type 2 diabetes, heart disease, allergies, gout, gastrointestinal troubles, skin problems, etc. Food also has social and political implications. Turtle soup, once a staple in places such as Mexico, is now considered inappropriate based on the scarcity of turtles. There is also the dynamic of social pressure with regard to choosing a diet. Some foods are considered to be especially healthy, like kale, lean meats, and organic fruits, to name a few, and we've all seen the parade of diets go in and out of fashion, such as keto, paleo, vegetarian and vegan, and raw foods. Then there is the question of sustainability. Is the food product sustainable, healthful, economical, tasty, popular, grown close to home, or grown far away? Do you subscribe to a Community Supported Agriculture (CSA) service or visit your Farmers' Market regularly, or do you eat fast food? All of these factors come into play when determining what, where, and how to eat.

What will food production look like in the future, next year, in five years, 10 years, 50 years, and beyond? What geographies, countries, industries, and companies will benefit, and which ones will struggle to produce food? How we grow and create food will change. We will utilize different growing techniques, alternative places of production, and create new distribution channels.

The future of food is exciting, complicated and hopefully will continue to be a satisfying experience. The food culture is getting more press and more exposure as backgrounds for major motion pictures, and television. Food and dining enjoy a plethora of online and television exposure; food and cooking programming have exploded on YouTube. Eating is something we generally do every day, it's a big part of our lives, and it's often enjoyable. In 2019, food was a monumental $1.1 trillion dollar business, according to USDA.gov. Now, more than ever, food technology is gaining massive traction. The future of food centers around the following possible investment opportunities, to name just a few:

Green agriculture

Dry farming

Hydroponics

Digital irrigation

Genetically modified crops and livestock

# Chapter 4

# WATER

**The following are a couple of stories that
made water up-close and personal to my life.**

In 2015, I met with a friend, Steve Carlson, in a pub in Redwood City, California for lunch. Now an intellectual property lawyer, Steve is a former Peace Corp volunteer who did a two-year stint in the 1990s in a small village in the high Atlas Mountains of Morocco. Steve and I are both members of Rotary International, a global service club that does philanthropic projects both at home and abroad. Steve spent a couple of years in the Berber village of Ait Daoud, Morocco in which about 1500 people live. Ait Daoud means the "the people of David." Morocco is a complicated country with a mosaic of cultural influences. Berbers are decendants of pre-Arab inhabitants of North Africa. They are scattered across Morocco, Algeria, Tunisia, Libya, Egypt, Mali, Niger, and Mauritania. That day in the pub, Steve pitched a project to bring year-round water to Ait Daoud.

There is a water source in the Atlas Mountains that, through gravity, brings water to Ait Daoud to serve the village's agricultural needs. It's a couple of miles from the village, and only delivers water intermittently. In the dryer season, the water dissipates in the earth and doesn't reach the village. Steve thought the construction of a concrete aqueduct to the village providing year-round water would be a

**Ait Daoud 2020**                              Photo contributed by Steve Carlson

good project for us to collaborate on and would provide much needed water and economic stability to the area. I agreed to help Steve and we worked together to build the villagers of Ait Daoud a better "ditch." By 2020, through relentless fundraising and after a few trips to Morocco, we funded and organized the building of an irrigation ditch lined with cement, to make a proper aqueduct for the village. This will increase the agricultural productivity for the villagers in a meaningful way. Once scarce and wasted into the dirt, water now flows to the village.

My experience with the Ait Daoud project brought me a deeper understanding of the meaning of water to communities. In the U.S., and especially in the affluent Bay Area of California, it's hard to fathom not having water at our fingertips. Paradoxically, the biggest areas in the world without fresh water are our oceans.

The challenge of sailing a small sailboat over a large ocean is akin to the challenges we experience as the population of our planet. On my first race from the West Coast of the United States to Hawaii as the skipper, I had to plan how much electricity we would use, how much food we would eat, and how much water we would need. We had to have adequate water to consume and additional contingency water in case there was an emergency, so when we arrived in Hawaii, we had the required excess, assuming that an emergency did not happen. The freshwater tanks on sailboats are notoriously unreliable, so I jammed bottles of water in every nook and cranny of the boat, under the floorboards, behind the cabinets, and next to the battery bank. We had enough water for each crew member to drink two liters a day, not for the 12-day minimum estimated trip length, but for a month at sea. We also filled the onboard tank for cooking and washing. Additionally, we had a system to turn seawater into fresh water that used reverse osmosis. It was a crude system that required using a hand pump to force seawater through fine membranes and would create about one gallon of fresh water in an hour of pumping.

The oceans are the largest deserts on earth. It's scary to be out there without adequate water reserves. The fear is very similar to that of the earth decreasing its fresh water supply based on global warming, evaporation, and ice melting.

### A Brief History of Water

Most scientists agree that the atmosphere and the oceans accumulated gradually over millions and millions of years with the continual degassing of the

Earth's interior. According to this theory, the oceans formed from the escape of water vapor and other gases from the molten rocks of the Earth to the atmosphere surrounding the cooling planet. After the Earth's surface had cooled to a temperature below the boiling point of water, rain began to fall—and continued to fall for centuries. As the water drained into the great hollows in the Earth's surface, the primeval ocean came into existence. The forces of gravity prevented the water from leaving the planet (NOAA).

Water has a profound physical, emotional, and economic impact on us as individuals on every part of the globe. For many of us, our first thoughts about water are personal. The introduction likely came from our parents through the importance of hygiene: water is integral to bathing and brushing one's teeth. As children mature and their horizons widen beyond themselves, they become aware of the importance of water for growing crops and raising livestock.

Water is the body's principal component and probably the most important resource to humanity. According to the Mayo Clinic, 60% of the human body is water. It is generally agreed that we, as humans, need about two liters a day to drink, and can only survive a few days without water; we can go about three weeks without food. We use about 80-100 gallons per person per day according to the USGS. That's a lot of water.

**Water Problems: Toxicity, Scarcity**

From the dawn of civilization, societies recognized the life-sustaining role of water. That awareness increased with the shift from nomadic to agricultural cultures. From a Judeo-Christian perspective, the Old Testament's Book of Isaiah quotes God as saying, "For I will pour water on the thirsty land, and streams on the dry ground ..."

Beer started as a safe alternative to water, since water was often tainted. Beer in ancient Egypt was actually a form of currency. In her book "Drinking in America," Susan Cheever documented that the Mayflower pilgrims decided to dock in Cape Cod, Massachusetts because they had run out of beer. While establishing what would become the most powerful economy in the world, they continued with beer as the primary beverage. In many cultures, tea was also a common option to water. People hadn't figured out the boiling of the water was actually the factor that made tea a healthy alternative to untreated water.

The peril associated with consuming tainted water remains a life-threatening one. According to "Water Wars: Privatization, Pollution, and Profit," polluted water is killing more people globally than any other single cause. Many factors contribute to the toxicity of water. The major ones are mining, earthquakes, extreme weather, and the general increasing aridness of the planet. In some communities, such as Flint, Michigan, lead has also made municipal tap water toxic.

In addition to toxicity, civilizations have also faced water scarcity. When California was being colonized by the Spanish in the early 1800s, water was scarce. Fresh water supplies were still tight during the gold rush in 1849. The population of California in 1850 was 92,597; now it's over 38,000,000 (www.census.gov, 2019). Natural sources could only support a tiny fraction of today's population. It took a monumental effort to get water to California's northern and southern populations from the Sierra Nevada mountain range. Since then, technology has been used to get water from the High Sierra for storage in reservoirs. In 1939, the 242-mile Colorado River Aqueduct was completed, bringing water from the Colorado River to the Los Angeles area. Consisting of more than 90 miles of tunnels, nearly 55 miles of cut-and-cover conduit, almost 30 miles of siphons, and five pumping stations and supplying approximately 1.2 million acre-feet of water a year - more than a billion gallons a day - the Colorado River Aqueduct helped make possible the phenomenal growth of Los Angeles, San Diego, and surrounding Southern California areas in the second half of the 20th century (ASCE.org). Nevada, Arizona, New Mexico, Colorado, Utah, and Wyoming also depend on the Colorado River for water. This demand has resulted in ongoing water wars among the seven states. Today there are more than 400 miles of aqueducts bringing water to California's population and the state's agricultural community.

Without these water diversion systems, California couldn't support its population or its status as the world's 5th largest economy. Without the current water industry, California would be a tiny shadow of what it is today: there would be no Hollywood, no agribusiness, no aerospace industry, no Silicon Valley, no tourism, and no commercial wine vineyards. This scarcity has positioned those needing water and not having an adequate supply, such as agribusiness and urban communities, as rivals.

Though at the time of this writing, recent rain and snow have partially mitigated the situation, California has been on the edge of a water catastrophe: it's been a well-publicized crisis. In many instances, we cannot get water at a restaurant in California unless we ask for it. Besides the use of traditional above-ground resources such as natural runoff, rivers, lakes, streams and reservoirs, water is also taken from underground sources. These underground sources have been so over-used in reaction to years of drought that there is little underground water left, causing agricultural businesses rethink the way they farm. Almonds, for example, use a lot of water when farmed in traditional ways but they can also be grown using dry farming techniques. Almond growers in California using traditional "wet" techniques will be more challenged, while dry farmers may have an easier time adapting.

## The Business of Water

Anyone who has ever seen the movie Chinatown knows that water is serious business and even —at least fictionally— a motive for murder. Set in Los Angeles in the early 20th century, Chinatown is about droughts, illegal water diversion, and business profits. Unfortunately, not much has changed since then. Water problems in California still exist and are accelerating.

According to the Oxford Dictionary, the term "rival" comes from the Latin term "rivalis," that is, "of the same brook" or "rivus." In some circles the meaning has evolved to denote those who use the same stream or those who are on the opposite side of the stream or river. The origin of this word suggests that water has always been a controversial commodity. There is a long-term social construct that water should be available and free to all, but that's not how it has been, and likely not how it will be in the future. Many believe that companies that profit from necessary drinking water are ethically and often legally challenged.

Bottled water ranked number 1 in 2019 as the largest selling packaged beverage (Rodwan, 2020). Companies that manufacture, distribute and sell bottled water employed over 137,000 people in the U.S. in 2020, with a payroll of over $6.3 billion. In 2020, the bottled water industry in the U.S. accounted for about $102.3 billion in revenue, slightly below 1% GDP (bottledwater.org). Water infrastructure is the other big player in the water business, accounting for $220 billion annually and employing over 1 million people (thevalueofwater.org).

In certain regions, water can create profit centers through its use for energy production, transportation, recreation, upscale real estate development, and tourism. When it forms a hub between two commercial areas, it can create wealth for both: the Hudson River, for example, which flows between Jersey City, New Jersey and Manhattan, New York. But the idea of water as an economic workhorse and platform for leisure did not always exist.

Even in the pre-scientific era, primitive societies were able to discern patterns in the variables linked to water. At the top of the list were the when, where, and how much of precipitation, both rain and snow; runoff of mountain meltwater; polluting elements entering natural water bodies, and extreme weather events. That observational data brought some degree of predictability. Societies could plan ahead, for example, which crops to plant for the next five years and when to leave the village because of a coming drought.

That was then. More recently, climate change is triggering significant, ongoing disruption in all aspects of the water cycle, or the hydrologic system. In some regions of the world, that has made both short- and long-term planning impossible. In other locations, even the most respected experts in fields ranging from weather to economics acknowledge the high degree of uncertainty brought about by this disruption. In terms of macroeconomics, this matters a lot. Paul Dickinson, chief executive of the UK CDP Water Disclosure Project, provides an insightful analogy. Dickinson points out that if climate change were a shark, then it's necessary to consider water its teeth. Disruption to Earth's water supply is potentially the biggest disruption caused by climate change.

The massive disruption in the worldwide water cycle can represent unique opportunity for investors. To recognize opportunities here, it is necessary to understand the moving parts in the hydrologic system, how they interact with each other, and the overall impacts on every aspect of the economy.

**Uses of Water**

Only 10% of available fresh water is used in personal human activities such as drinking and bathing. The rest has become a necessary part of business. That means the current scarcity is a serious threat to business and, therefore, global GDP growth. That is what is putting climate change on the radar of business.

Throughout the world, the grim memory of the global economic crash of 2008 hovers over the business sector. As media coverage and marketing messages indicate, there is an intense preoccupation with which nations are currently in recession or will soon be. Speculation then focuses on severity and duration. It will take several generations for the 2008 economic downturn to be integrated; until then, anxiety will be embedded in business. That hyper-alertness to any internal or external threats to an enterprise's success or turnaround since 2008 has come to involve climate change in general and water availability in particular. There might not be an agreement about what causes climate change, but denial of its existence and possible future impacts is lessening. Those in the business sector who recognize the water scarcity problem likely realize it is going to get worse. The question is: how do we adapt to the decrease in this resource?

There has been magical thinking that once a society realizes the realities of climate change, the situation will go away. Fortunately, that undue optimism is lessening. The new objective has become mitigation, a mindset and behaviors aimed at reducing the severity of the situation, treating the pain and, hopefully, preventing a worsening in the future. For some businesses, such as agriculture, the threat of water scarcity is worse than for others. But all business sectors have to re-think how to manage this resource. Some are leading the way by inventing technologies, products and services to mitigate the constraints imposed by water scarcity, even figuring out ways to monetize their approaches. These technologies include rainwater harvesting, drip irrigation, and desalination.

The impact of increasing water scarcity on the business sector could also trigger panic, among both business owners and government. The "rival" ethos could play out in everything from diversion of supply to litigation. Government entities could impede reforms on behalf of environmentalism, worsening the impacts of climate change. In addition, an economic downturn or downright collapse triggered by water issues could topple political leadership. The new order could be pro-or anti-environmentalism. These are possible scenarios investors have to consider. Wealth can often be created out of business failure, changes in governmental policies, and political coups.

In fact, there is less fresh water today than ever before, because of the increase in global population and higher temperatures of climate change which cause evaporation. Fresh water accounts for only 2.5% of the earth's water. According

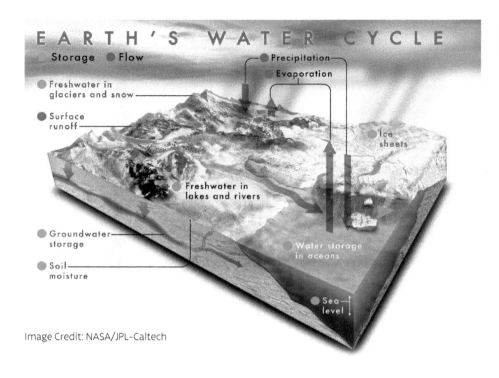

EARTH'S WATER CYCLE

Storage ● Flow — Precipitation —
Evaporation
● Freshwater in
glaciers and snow —
● Surface
runoff —
Ice
sheets
Freshwater in
lakes and rivers
● Groundwater —
storage
● Soil —
moisture
Water storage
in oceans
Sea —
level
Image Credit: NASA/JPL-Caltech

to National Geographic, of the 2.5%, only 1% is easily accessible (2020). The rest is trapped or — depending on your point of view— stored in glaciers and snowfall. Yet, the danger is that if glaciers and snow melt too rapidly because of higher temperatures, there could be flooding and rising sea levels. As a natural form of irrigation, that water is gone forever once the glacier melts. Because it evaporates so quickly once it becomes a liquid, it will not be available for the future. Society will have to engineer artificial forms of irrigation, shift from the current industries, or migrate.

But the problem isn't as simple as sea level rise and droughts. Our assumptions about the availability of water as described by the water cycle we learned in elementary school are no longer reliable, because the water cycle has been disturbed by climate change. According to Dr. Marina Oster, PhD, climate scientist and Stanford University lecturer, the key to understanding how the water cycle has been altered is understanding precipitation recycling ratio. Precipitation recycling ratio describes how the amount of water that has evaporated from a given area contributes to precipitation over that same area. Changes in our climate have resulted in a dynamic whereby water is collected in storm systems, held, and then released in different locations, often resulting in extreme water events such as flooding and resulting in runoff lost to the oceans.

Scientists are presenting alarming data regarding our global water supplies. In June 2019, NASA Earth Science News Team member Carol Rasmussen published the article "The Water Future of Earth's Third Pole" in the NASA publication "Global Climate Change." She explains that Asia's high mountain ranges such as the Himalaya and Hindu Kush provide water for drinking and irrigation for more than a billion people. The water is delivered via meltdown from the snow and glaciers. That is why some refer to this region as the "Third Pole." But all is not well at the Third Pole. The water delivery system in this region is threatened by climate change. The good news is that the alert in nations such as Pakistan and China are adapting to the changes. They are shipping in water from elsewhere, using desalination plants, and building new power plants to move water to where it's needed. This is an example of a developing problem and an indication of a strategic response by some of those whose enterprises and lives are being impacted. This snapshot does not represent science-fiction time travel into the future: societies around the world are being threatened —and are adapting— right now.

Many places are currently suffering serious water shortages: Yemen, Libya, Jordan, Western Sahara, and Djibouti among them. These places are all in arid climates, most in sub-Saharan Africa. Because their economies are developing, they have a significantly lower margin of error than do developed nations. Other places struggle with tainted water sources: Mexico, Congo, Pakistan, Bhutan, Ghana, Nepal, Cambodia, Nigeria, Ethiopia, and Uganda among them. Polluted water has a profound effect on health and the economy. People get sick from bad water, putting a strain on already challenged health care systems. Less water limits crop production and ranching output, putting even more stress and downward pressure on these developing economies and further exacerbating poverty and mortality.

**Desalination and Filtration**

Desalination technology creates fresh water by separating salt out of water, producing drinkable water. This can be accomplished in a number of ways. The most common technology is reverse osmosis, which uses high pressure to drive saltwater through fine membranes. This technology is not new. It requires a lot of energy. Certain areas of the world are becoming increasingly arid, especially coastal areas that have a cold current which generally lessens rainfall. Those areas might have to eventually divert economic resources from other competing areas to pay for desalination. The major downside of this technology are the energy costs associated with the process: all energy consumption is being reviewed by scientists and those on the front lines of environmentalism. But with the desperate need for water growing worldwide, both business and government are exploring forms of reverse osmosis. In 2002, Singapore announced that a process named NEWater would be a significant part of its future water plans. It involves using reverse osmosis to treat domestic wastewater before discharging the water back into the reservoirs. Namibia is also using reclaimed water in response to its serious water shortage, employing a variety of techniques, including reverse osmosis.

Industries that might benefit from increased desalination activity are component suppliers, sub-manufacturers that make parts for the device suppliers, energy companies that supply the energy to power desalination plants, and companies that make the equipment for the energy suppliers. These energy companies could include those that use alternative energy sources such as wind, solar, and hydro-power. In this way, one impact of climate change — water scarcity— could benefit many companies and, ultimately, their shareholders. The leading players in the global reverse osmosis membrane market are:

Dow Chemical Company

Toray Industries, Inc.

LG Chem Ltd.

Nitto Denko Corporation (Hydranautics)

Toyobo Co., Ltd.

Lanxess AGKoch Membrane Systems, Inc.

Water filtration is another area to consider. As the planet becomes drier and water becomes scarce, what water remains in the streams, lakes, and water table may become contaminated and brackish, leading many more areas to struggle to find clean water. Certainly sub-Saharan Africa, parts of Central and South America, and parts of Asia will need to be more and more concerned about clean water. Currently numerous non-profit agencies are installing solar powered water filtration systems in remote areas. These systems cost tens of thousands of dollars. As time goes on, this type of system will likely be more and more in demand, and with larger missions. Though current systems are designed for villages with populations in the hundreds of people, over time, towns in the thousands, hundreds of thousands, and ultimately in the millions could be in need of water filtration where need hadn't existed before. Water could become "liquid gold."

## Profiting from Mitigation

The realities of the water crisis, both in commercial activity and in everyday human life, are generating what one might call the Mitigation Industry. On the one hand, this dynamic opens up fresh possibilities for doing business and investing. On the other hand, some solutions are themselves causing increasing environmental damage. Ironically, the result is further opportunity to build wealth through "fixing the fixes." The classic case is bottled water.

The horrors set in play by polluted drinking water are captured in the phrase Montezuma's Revenge, the old-line code for the intestinal distress many tourists endured after drinking local water in Mexico. Among other effects, it started the craze of drinking bottled water when traveling anywhere, a practice that has evolved into drinking bottled water even at home. Incidentally, Mexicans drink bottled water too. The Mexican government is still unable to guarantee the quality of tap water. In addition, the subpar distribution system of water, taxes on sodas, and awareness of the health benefits of water have made Mexico the largest consumer of bottled water in the world.

What is going on in Mexico is a microcosm of how to profit from mitigation. In turn, efforts at mitigation themselves create their own environmental problems, opening the door to other opportunities for wealth creation. The law of unintended consequences dominates solutions to climate change.

According to Euromonitor International, Mexico's bottled water industry in controlled by three companies: Danone (47%), Coca-Cola (19.4%), and Pepsico (7.1%). The bottled water business, just in Mexico, represents about $9 billion dollars annually. In some countries, it's the primary source of drinking water. The business is huge, its potential growth is almost infinite. Also, the global wellness industry emphasizes the health benefits of water. Overall, the wellness industry generates $4.2 trillion annually. Of that, the healthy eating segment, which includes recommendations for increased water consumption, produces almost $640 billion a year.

Water is big business. A host of companies have recognized this and capitalized on the increasing popularity of bottled water. Water may be a basic human need, but brand value and loyalty are clearly at work in the bottled water industry. As of this writing, the top ten bottled water brands in the world represent a global group of companies, all competing for a share in the water market. The top companies and their brands include:

Danone (France): Mizone, Aqua, Evian

hint (San Francisco, CA)

Nestlé (Switzerland): Perrier, S. Pellegrino, Poland Spring, Acqua Panna, Vittel

PepsiCo (Purchase, NY): Aquafina

The Coca Cola Company (Atlanta, GA): Glacéau smartwater and vitaminwater, DASANI, Kinley, Seagram's

AJE Group (Spain, Peru): Cielo

FIJI Water (Los Angeles, CA)

Hangzhou Wahaha Group (China): Wahaha Active Oxygenated Water, Wahaha Purified Drinking Water

Icelandic Water Holdings (Iceland): Icelandic Glacial

Mountain Valley Spring Company (Hot Springs, AR): Mountain Valley, Diamond Valley

Packaging innovation is an important part of brand identity, with manufacturers constantly introducing new formats to attract and delight consumers. In a sense, packaging constitutes entertainment. For bottled water products, packaging is undoubtedly important: it enhances shelf appeal, differentiates the product in the market, and consequently impacts sales. Today, bottled water manufacturers are increasingly choosing eco-friendly packaging formats and luxury formats, thanks to the rising concerns about the disposal of plastic bottles. National Geographic notes that estimates for how long it takes for plastic to decompose "range from 450 years to forever." Recent research by the University of Hawaii found that discarded plastics emit powerful greenhouse gases as they disintegrate (Royer, 2018). Heat accelerates the gas release, so the more temperatures rise in climate change, the more quickly gases will be put in the environment. The looping dynamic is obvious. If incinerated, plastic bottles set off toxic fumes.

In response to these negative environmental impacts, bottled water companies are introducing initiatives to promote eco-friendly packaging. For instance, in March 2017, Danone and Nestlé joined forces with California-based Origin Materials to form the NaturALL Bottle Alliance. Together, they aim to develop and launch a PET plastic bottle made from bio-based materials. Going green is a proven marketing strategy. The increasing packaging innovations in bottled water are expected to impact the growth of the global water market. Businesses making these kinds of adaptations will leverage them as forms of added value. It's important to remember, though, that plastics cannot be recycled indefinitely. According to Dr. Oster, about three plastic lives is the limit: the resin from a plastic water bottle can be turned into a laundry detergent container, then a speed-bump, before becoming garbage. There are opportunities and some hope on the horizon even at that point though, as companies are developing ways to keep plastics out of the garbage. One such company is Eco Fuel Technology, Inc., which converts plastic bottles into consumption-ready fuels.

## Smart Technology

Technology, from rudimentary to digital, has always played a role in the water cycle. Irrigation, the practice of transporting water from where it is to where it is needed, dates back to ancient times. In Egypt, in 3100 B.C., King Menes developed a system of diverting water in a controlled manner for intermittent release. In ancient Greece, one primary technology was organizing the system of human labor and routes. Those at the bottom of the food chain traveled long distances to collect and transport water. On the Greek islands, people constructed water troughs in high places to collect rain and control its release. Today, of course, consumers regularly have water transported for their personal needs.

But while irrigation delivers water, it also adds expense. In a time when the agriculture and livestock industries are experiencing reduced margins, there is constant research into how to protect profit by reducing irrigation costs. Smart technology may be the long-term solution. In a 2018 article, Amanda Shiffler explains how smart irrigation systems can both conserve water at the front end and increase profit margins (Shiffler, 2018). Shiffler notes that the best systems bundle a number of approaches, continually analyzing the key variables: soil condition, evaporation rates, weather, and plant consumption rates. She identifies ten companies whose smart irrigation systems stand out for their innovation:

Hortau

HydroPoint

Arable

Tule

Droplet

CropX

Tevatronic

AquaSpy

SmartFarm

Pow Wow Energy

**Water for Livestock**

Water is not only a challenge for irrigating crops: it takes a lot of water to facilitate maintenance of livestock. With water scarcity brought on by climate change, it could become more difficult and more expensive to operate the livestock industry in the current manner. That is why some enterprises developing alternatives to food from cows, for example, are experiencing commercial and stock market success. For example, following its well-received IPO, Beyond Meat's stock price continues on an upward trajectory. It's important to look at the math. According to National Geographic (Blogger, 2017), the water required to produce several common foods is approximately:

single chicken egg — 53 gallons

one pound of chicken — 468 gallons

one gallon of cow's milk — 880 gallons

one pound of beef — 1800 gallons

That includes water to grow the crops to feed the livestock, drinking water for livestock, and so on. Growing crops to feed animals consumes about 56% of water used just in the U.S.

## The Future of Drinking Water

Water is the elixir of life. It's one of the most important, if not the most important, resource on this planet, more important even than food. There is no life on earth without water. There has been human conflict over water since the beginning of time. Water is pivotal to our changing climate as it dramatically impacts our future, our survival, and the future of our economy. These are some of the investment opportunities related to new and disruptive water technologies.

Drinking water packaging

Desalination and associated technology

Irrigation technology

Water filtration technology

# Chapter 5

# TRANSPORTATION

Whether it's leisure travel or shipments of raw materials, the transportation industry is a big part of the economy and a key measure of economic activity. In 2014, transportation contributed to 8.9% of the U.S. economy, according to the Bureau of Economic Analysis. Transportation is the mechanism that allows economies and society to function, akin to blood flowing through our bodies. In ancient times it was on foot, mainly for hunting and gathering and occasionally for getting around to fight with or flee from other tribes. Then, over 6000 years ago, people began domesticating camels, oxen, horses, elephants, and other animals to expand their range and to carry more stuff.

I am a travel junkie. I've always enjoyed exploring, sometimes too much. When I was in high school, I did two major bicycle rides with my friends. At the age of 15, my friend Michael and I rode from San Francisco to Los Angeles, about 600 miles. It took us about a week. Then the following year, another buddy, Eric, and I had our bicycles put into boxes and into the belly of a DC-9 aircraft to be flown from San Francisco to Vancouver, Canada. The idea was that if we put our bikes on the plane, there was no turning back; we had to ride the 1200 miles to get home. We averaged about 80 miles a day. It was a slog, but we made it in about two weeks. We were kids, we had no outside support, no van following us, and no hotels to stay in at the end of each day. We had two cheap 10-speed bicycles worth under $100 each, with less than $100 in our pockets and no credit cards. We ate cheaply and slept on the ground when it got dark, off the side of the road. These two bicycling efforts were my first real forays into crazy unsupervised adventure.

I continue to ride bicycles, but recently, I've been exploring the world on motorcycles: it's less work than pedaling and often more exciting, but sadly somehow less rewarding. In 2019, I re-lived my San Francisco to Los Angeles bicycle ride, this time on my Ducati Monster motorcycle. It was fun to ride the same route on Highway 1, the Coastal Route or Pacific Coast Highway. As a sailor, I have traversed many of the world's oceans including the Atlantic, Pacific, Southern Ocean, Arctic, and Antarctica. Sailing is the ultimate low-impact transportation model. In theory, one could sail indefinitely using only the wind; the reality in the 21st century is that we need generators or solar panels to maintain communications and keep the WiFi going. As a professional pilot, I've seen a lot of the planet from the air, transiting the continental U.S., Hawaii, Europe, Mexico, the Caribbean, and the Middle East. I've flown totally around the world a couple of times: it nev-

**Headed from San Francisco to Hawaii 2012**                                    Michael Day

er gets old. But the way we get around the world is changing. There are systemic disruptions to the way we travel on the roads, at airports, on trains, and at sea: constant construction, security concerns, pollution, and energy limitations.

Seagoing vessels have been around for at least 10,000 years in Europe and in the Middle and Far East. The Chinese and ancient Greeks had been sailing for a long time for trade, basic transportation, war, hunting fish and whales, and for hauling passengers and freight. Trains became a major force in the early 1800s. Automobiles and motorcycles became fashionable in the early 1900s. Trains, cars, and trucks fueled the industrial revolution. Airplanes became useful in the 1920s, and jet planes changed the airline and military-industrial complex in the 1950s. In the 1960s space travel emerged, propelling humans into orbit and to the moon in 1969. In the 1970s unmanned space travel to other planets began. And in the 1980s, the space shuttle acted as a delivery truck for satellites, astronauts, and equipment to build and maintain the International Space Station.

The 1990s began a race for transportation vehicles of all types to become more fuel efficient, less expensive, and more accessible. Sadly, large auto manufacturers spent decades since the 1900s buying up patents for fuel efficient automotive technology, especially electric cars, and withholding those patents to prevent competition.

In the 2000s electric vehicles (EV), autonomous vehicles, and private space companies have become the norm. The future of transportation has recently become a phenomenon of entrepreneurs such as Elon Musk, the father of both Tesla and SpaceX. Tesla and SpaceX were laughable to the mainstream transportation industrial complex — until they weren't. Just recently, SpaceX has taken over NASA's job of delivering goods and people to the International Space Station.

Now most of the auto manufacturers are in the midst of reinventing themselves as lean, green fighting machines. Tesla, Nissan, Audi, Chevrolet, BMW, Jaguar, Porsche, Kia, Mini, Hyundai, Toyota and others are seriously engaged in the EV automobile market. Some, such as Tesla and Toyota, are doing better than others. Toyota revolutionized the hybrid market by combining gas engines with electric motors. In fact, Toyota was so successful that their hybrid engine technology was used by almost all of the competition in the early years of hybrids.

Besides personally owning two hybrid automobiles and a Jeep that runs on sustainable biomass fuel, I recently traded my gas-guzzling Ducati Monster motorcycle in for a state-of-the-art electric motorcycle. The motorcycle is roughly the same size and shape of my old Ducati sport bike, but the new Zero is virtually silent, uses zero fuel, and is powerful beyond belief. Even the famous old-school Harley-Davidson motorcycle company is entering the electric motorcycle market, a big departure from its petroleum burning "tough guy" cultural background. Harley's EV motorcycle, the LiveWire, has similar performance to the Zero motorcycle.

Another seemingly unconventional but extremely popular addition into the transportation world is e-bikes. These are electric powered or electrically assisted bicycles. I rode my first e-bike in Copenhagen: it was fantastic. The bikes were inexpensive to rent and available just about everywhere. I still had to pedal but it was as if I had superpowers: I pedaled at a moderate pace, and the bike went at a pretty good clip. It seemed impossible to get tired. No matter the terrain, I kept my pace. What wonderful tech! Some e-bikes are like motorcycles with throttles or combinations of pedaling and/or throttles. These bikes have become extremely popular, especially during the COVID-19 crisis.

**My Zero Motorcycle** Photo Michael Day

While there is clearly great potential for EVs and ebikes, the charging infrastructure required remains a challenge — and a potentially burgeoning industry. In the U.S., the State of California has taken the lead with plans to invest $1 billion in charging facilities. California also has a plan to ban gas-fueled cars by 2035. On its face, this is a thoughtful proposal, but the problem is the recharging. I have been operating an electric vehicle for some time now, and while I can plan a trip based on recharging stations along my route of my travels, sometimes it just doesn't work out. The problem is if for example, an electric vehicle might have a 250-mile range and the destination is 300 miles away, one would have to stop to recharge once along the way. With fast charging, the re-charge can conceivably be performed in less than hour, just the right time for lunch, but there are two problems I have experienced: the charging station can be full with a long wait of cars in line at the charging location, and/or some or all of the charging stations are inoperative. To make matters worse, many of the charging stations I have been to allow the vehicles to stay at the charging station for up to 4 hours. This situation can lead to a several-hour delay in a planned trip. Most of the time, I have been the only one at the charging facility, but I have had experienced lines or "out of service" charging stations, leaving me unable to charge my vehicle. Fortunately, I always plan my trips on my electric vehicle so I can make it round-trip without absolutely having to recharge. I use a non-electric or hybrid vehicle for longer trips. With California's current population of over 39 million, with the goal of operating predominantly electric cars by 2035, a high demand would be put on recharging stations which could add hours to a planned trip is everything doesn't work out just right. In 2019, 7.7% of cars in California were plug-in electric models. In 2035, perhaps 90% of California cars could be plug-in. I would suggest improving hybrid technology allowing the use of electric vehicle's range to be extended by use of a motor of some type; perhaps fueled by hydrogen. The outcome would be using recharging stations as an option, but not as an absolute necessity.

Most people believe that battery-powered cars are the future. But Hyundai, the South Korean vehicle maker, differs. Hyundai has been running a worldwide campaign promoting the benefits of an alternative to electrical power—hydrogen fuel cells. Instead of storing and then using electricity from batteries, fuel cells generate current from a chemical reaction between hydrogen and oxygen. The oxygen comes from the ambient air; the hydrogen is compressed and stored in a tank on the vehicle, and can be refilled at a hydrogen station. Hydrogen fuel cells don't create exhaust the way traditional engines do: the exhaust is simply water, the result of the reaction of the hydrogen and oxygen. While certainly a challenge, the hydrogen refill infrastructure is being addressed globally. According to the H2tools.org website, hydrogen fueling stations are being expanded in 33 countries. California is building a plan for hydrogen fueling stations as a result of a 2013 bill that was passed.

In the railway sector, electric and magnetic trains are the future. Many light rail and local trains are being converted to more efficient electric power. The real breakthrough is magnetic wheelless trains that float on the air. The new Chinese maglev trains utilize magnets to float their carriages above the ground. These are the fastest form of rail travel in existence. The Shanghai maglev train that connects Pudong International Airport to a major metro terminal outside the city is currently the fastest. The 19-mile journey takes 7 minutes to complete at speeds of nearly 270 mph.

**Ships at Sea**

Oceangoing vessels are the backbone of international trade and commerce. You can't send raw steel from China to Japan in a truck or on an airplane. Heavy, big, and often non-urgent freight is sent around the world via big container ships. With a purchase price in the range of half a billion dollars, these vessels weigh up to 220,000 tons or 440 million pounds and are 440 meters or 1,443 feet long. Bigger is better in container ships as it's more efficient to operate bigger ships, as long as they aren't too big to dock at major ports.

These big ships burn a lot of fuel and cost a lot of money to operate. Fuel consumption on ocean freighters or containerships is mostly about ship size and speed. An average ship could burn about 225 tons of fuel per day at 24 knots. At 21 knots, this consumption could drop about 33% to 150 tons per day. Less speed equals less fuel but, while it's more economical to go slower, it's not always practical to do so. In 2018, the International Maritime Organization (IMO) tentatively agreed to make new ships more efficient in an effort to implement the organization's initial greenhouse gas (GHG) policy. To reduce emissions of greenhouse gases from maritime shipping, the energy efficiency design index (EEDI) has been made mandatory for all new ships. Steps have been taken in the form of new technologies features to ensure that the EEDI is met (Mahendra Singh | In: Marine Technology | Last Updated on October 13, 2019).

Small vessels with shorter missions are embracing electric propulsion. While most water vessels are powered by diesel engines, with sail power and gasoline engines also popular, boats powered by electricity have been used for over 120 years. Electric boats were used from 1880s through the 1920s, when the internal combustion engine became dominant. The following companies make electric motors for boats:

ABB Marine

Aquamot

Combi Outboards

Elco Motor Yachts

ENAG

Fnm Marine

GreenStar Marine

Hoyer Motors

Where are we headed globally in the area of transportation? Certainly, COVID-19 and the inevitable aftermath will change the mass transportation industry in substantive ways. People are hesitant to get on airplanes, buses, trains, and cruise ships. Since various pandemics such as SARS, H1N1, and COVID-19 will likely continue to plague civilization every few years, the transportation industry will continue to be susceptible to economic strife. Global warming is also affecting the way we move around. People's concern over polluting the atmosphere with hydrocarbons generated by traditional forms of transportation leads some to reduce their carbon footprint by flying and driving less. Certainly, Webex, Zoom, and other web conferencing tools are going to take a piece out of the airline industry by reducing businesses' willingness to fly employees around. There will likely be two main impacts on the industry's income. First, fewer people will fly due to economics and environmental concerns. However, while pandemics and environmental concerns may throw a wrench in the works of airline passenger demand, the gross overall numbers of passengers will likely increase over time. In fact, global passenger outlooks for air travel have been forecasted to skyrocket in the next few years. The second impact will likely be an even bigger push to lower costs in airline travel from building cheaper more fuel-efficient planes, to reducing crew and logistical costs.

The investment opportunities in sea and land transportation are many. They include boats, cars, motorcycles, trucks, trains, power plant manufacturers, sub-suppliers, and recreational vehicles. Other big opportunities include infrastructure, logistics, and support. As time goes on with more severe weather, limitations on carbon usage, and changing population centers, the way we get around and move supplies will change. We can expect more fuel-efficient vehicles, rightsizing of fleets, and re-rationalizing supply chains and frequency of transport.

It appears that the future of transportation lies in electric vehicles, energy sources, and infrastructure. Electric cars accounted for about 2.1 million vehicles or 2.6% of global auto sales in 2019 (International Energy Agency, Global EV Outlook 2020). By 2025, JP Morgan estimates electric and hybrid vehicles will account for 30% of all vehicle sales. The EV industry is changing rapidly in front of our eyes, creating an insurmountable disruption away from petrochemical vehicle production towards sustainable vehicle production with specific opportunities in the EV and battery production sectors. The costs for battery technology for electric vehicles have dramatically reduced over the years. Taking advantage of this cost benefit, Chinese companies are taking the lead in the production of electric cars and their components, accounting for an estimated 59% of the market by year end 2020. Chinese companies AIC, BYD and ZHIDou are the major players. California's (soon to be Texas's) Tesla, specializing in EV and battery technology, is also a major force in this disruptive new technology.

The transportation sector is on the precipice of a revolution in terms of propulsion, vehicle structure, and AI that drives the internal technology. The investment opportunities come from many directions including vehicles, powerplant technology, and infrastructure. One obvious infrastructure consideration is electric car refueling stations. Tesla touts over 20,000 global superchargers, a number that's growing daily. These stations can recharge Tesla vehicles for up to 200 miles of range in 15 minutes. Another less obvious element of transportation involves the logistical technologies that surround air traffic control, especially integrating newer technologies like flying cars and drones into the existing flow of airline, corporate, military, and private air travel. Transportation as a sector is ripe for investment exploitation, especially in the sustainable space.

The following are macro investment opportunities in the transportation arena:

Electric Trains

EV Automobiles

EV Motorcycles

EV Bicycles

Electric Motors

Batteries

Vehicle Parts

Transportation related Technology

Transportation AI

# Chapter 6

## AVIATION AND SPACE

Civil aviation accounts for $1.8 trillion in economic activity and $488 billion in earnings. The industry generates 10.9 million jobs and contributes 5.2% of GDP in the U.S. (FAA.gov). In 2018, the FAA estimated the U.S. space industry was valued at approximately $158 billion in 2016. In 2018, the U.S. aerospace and defense industry (A&D) added more than $374 billion to the GDP of the U.S., totaling 1.8% of U.S. GDP (www.aia-aerospace.org/2019-facts-and-figures).

Mankind had dreamed of flying since time began. At the turn of the last century, powered flight became a reality. On December 17, 1903, the Wright brothers flew a heavier-than-air aircraft, the Wright Flyer, 4 miles south of Kitty Hawk, North Carolina. These two crazy bicycle mechanics made history that day, ushering in a new era in transportation with their disruptive technology. Rudimentary airplanes were used in World War I to deliver crude bombs, fire guns, and perform observation flights behind enemy lines to see what the other side was up to. It took another decade for airplanes to begin dusting crops and delivering mail, cargo, and passengers. These airplanes were a giant breakthrough.

**The Airlines**

Technically DELAG, Deutsche Luftschiffahrts-Aktiengesellschaft, was the world's first airline. It was founded in 1909 in Frankfurt, Germany and operated airships manufactured by the Zeppelin Corporation. Contemporary airlines came shortly after: KLM began in 1919, Qantas Airlines started in 1920, Aeroflot in 1923, and American Airlines in 1926. From the early 1920s to today, it's questionable whether passenger airlines ever made any money. If we measure all of the cumulative expenses against the cumulative income (before and after regulation and subsequent deregulation of the industry), the profit of the passenger airline industry would be lackluster at best. In my opinion, the cyclical airline industry is akin to a public utility, a commodity business that suffers big swings in profitability with thin profit margins. This cyclical business sector experiences risk from all directions, and I don't see any systemic change to this dynamic any time soon, if ever. However, this doesn't mean the aviation sector as a whole does not present opportunities.

**Aircraft Manufacturing**

Once a very competitive industry made up of dozens of manufacturers, the airline manufacturing business has dwindled down to basically two, Boeing and Airbus. The light aircraft and corporate aircraft industries thrived after people and organizations turned to private aircraft as security at big airports became onerous after the terrorist attacks of 2001. Since a spike in sales in 2008, general aviation aircraft sales have declined. It's my opinion that new small planes cost too much. At the time of writing, an entry level plane such as the four-seater Cessna 172 costs about $300,000; a six-seater light twin-engine aircraft such as the Beechcraft Baron costs about $1.4 million; and a medium-sized business jet with 7 seats, the Cessna Citation X, costs about $23 million. The average cost of a private pilot license is about $10,000. The whole small airplane world is expensive with bad economies of scale. The demand persists, though, as aircraft are still being built for personal and business use by the following manufacturers:

## Drones

Drones have been a great, useful, and economical disruptive technology. Drones are used for military, law enforcement, search and rescue, filmmaking, news reporting, crop dusting, recreation, surveillance, firefighting, remote sensing — the list goes on. This is really a solid industry with many potential uses and associated benefits of related technology such as the photographic and sensing equipment that can be carried on board. Even emergency medical equipment and medications can be flown into remote locations with drones. Top drone manufacturers include:

| | |
|---|---|
| DJI | Hubsan |
| Yuneec | Parrot |
| UVify | |

## Flying Cars

Flying cars are a not so new technology that is being closely scrutinized. They have been around for a long time but never really got off the ground. The first serious version of a flying car was the Airphibian in 1950, which was a plane adapted to drive on the road. As a pilot, I have a lot of questions and concerns about flying cars. The crux of my concerns lies with safety. In August of 2020, according to the Robb Report, New Hampshire became the first state to legalize flying cars. To fly such a car requires a pilot's license, and takeoffs and landing can only be performed from airports. There are other flying car aircraft that are being built, but most fall into the very limiting FAA Ultralight category (Ultralight craft are not allowed flight in urban areas or at night). The new flying car technology revolves around manned drones. Technologically, it is difficult to define the difference between manned and unmanned drone technology. For the tech to work, flying cars with passengers need to be highly automated, very similar to a drone. This technology is very plausible and useable although there are big problems surrounding keeping them from crashing into each other and other objects. It's my opinion that new technology has to be developed to create dedicated pathways for these aircraft to operate. There must be foolproof and interactive communications between these aircraft to avoid collisions. This technology exists today, but it currently requires manual intervention, therefore the technology needs to become automated. The devil is in the details, in the case of flying cars, and there is a long list of critical details that must be addressed. Just because we can make and fly flying cars does not mean we should, and definitely not before the infrastructure is in place and well tested. This will require a monumental effort between the manufacturers, the FAA, the Department of Homeland Security, and federal, state, and local governments. It will be a logistical nightmare to really have useful flying cars in the near future.

There are major investment opportunities in aviation. The opportunities are more subtle than the advent of supersonic and hypersonic technologies. Supersonic is defined as flying faster than the speed of sound, while hypersonic is generally defined as flying greater than 5 times the speed of sound. The new frontier in aviation surrounds tapping into better efficiencies such as more fuel-efficient engines, better and more effective airframe designs that can carry more weight with less fuel, and better computational capabilities to further maximize efficiencies. The computational efficiencies can improve the air traffic control systems, scheduling systems, pricing, and other operational efficiencies. The technological advances, while not always breathtaking, can be endless small improvements to

**Futuristic Flying Car**

Pavel Chagochkin; Shutterstock

improve operation and capabilities while keeping costs down. Opportunities in aviation include the following areas:

Airplane manufacturers, large ones

Drones

Jet engine manufacturers

Logistics, Support, and Airport Facilities (these are the services that move goods and services around and keep the airports functioning, such as catering, fueling, passenger ground transportation, and security).

## Space Travel and Space Science

When I was a kid, I had my face stuck to the television set often, but never as intensely as July 24, 1969, when all three television networks showed NASA astronauts Neil Armstrong and Buzz Aldrin landing safely on the moon. It was one of those few times that, even as a kid, I knew instantly that history was happening, at that moment. The Eagle spacecraft landed on the moon and subsequently lifted off to return Neil, Buzz, and Command Module Pilot Mike Collins back to earth safely and triumphantly.

Fast forward 50 years to mid-2020, when NASA astronauts Robert Behnken and Douglas Hurley were the first to be transported by a private U.S. space company on SpaceX's Crew Dragon capsule. The mission successfully delivered the two astronauts to the International Space Station on May 30, 2020 and returned them safely to Earth on August 5, 2020. This was also a history-making event, but a different kind of historical breakthrough: it was a business first. NASA had outsourced its transportation of astronauts into space to U.S.-based private enterprise.

As a teenager, I had visions of a career as an astronaut in the Space Program at NASA. I did a short stint at Ford Aerospace as a satellite technician during college. While at Ford, the concept of building, testing, and participating in the launching of spacecraft seemed cool, but the whole job was pretty boring, with short episodes of being even more boring. I learned that engineering delay after engineering delay created too much down time, especially for a 19-year-old. (An "engineering delay" was code for something had gone wrong and no one was sure how to fix it.) As a junior in college, I suffered from a fatal bout of pragmatism, and made the decision to pursue a career as an airline pilot instead of an astronaut. I figured I had a much better chance of getting an airline job versus an astronaut job. Later in my airline career I had the opportunity to meet a few Apollo astronauts on my flights, mostly to Washington, D.C. One of my flight instructors at TWA was an Apollo astronaut and an X-15 pilot. The guys I met were all smart, humble, and very accomplished. My wife and I attended the Explorers Club annual dinner in New York City at which Buzz Aldrin was an emcee. We had the honor of briefly meeting Buzz. Though in his late 70s, he was very engaging and heavily involved in planning a manned Mars mission. He was the epitome of a confident and accomplished explorer.

The heady science fiction of space exploration seen in Star Trek and Star Wars hasn't come to fruition, but we have come a long way as a society integrating space technology into everyday life. Global Positioning System (GPS) technology is such an integral part of our world today. The system is used for navigating cars, buses, planes, and bicycles: that's the obvious stuff. The more subtle uses include wearable fitness devices and safety equipment such as motorcycle airbag jackets that deploy when the GPS inputs and accelerometers sense sudden stopping. GPS is used to track lost dogs and stolen cars. The list goes on and on. We've become so dependent on GPS that, should the system become degraded by outside forces, airliners in flight go into emergency navigation modes limiting landing capabilities and general navigation. GPS is no longer a luxury, it has become basic to our business, commerce and lifestyles. We may not be traveling to Alpha Centauri for vacation, but we're using GPS to get to Yellowstone National Park.

When I was working at Ford Aerospace, we built telecommunications satellites that provided voice and data capabilities from Earth orbit; I worked specifically on Intelsat. Today's Intelsat website says: "...We combine the world's largest satellite backbone with terrestrial infrastructure, managed services and an open, interoperable architecture to deliver high-quality, cost-effective video and broadband services anywhere in the world..." Satellite TV has become huge to deliver television to homes and businesses in both remote and urban locations. Satellites bring internet and telephone connectivity to places that had no such service in the past.

I had the privilege of going to Antarctica a dozen or so years ago on a raggedy old Russian boat on its last legs as part of an eco-tour from Ushuaia, Argentina to the Weddell peninsula in Antarctica. Even though the boat was built circa 1959, it had satellite communications capabilities that allowed us to communicate with our families during our very rough crossing from South America to Antarctica. Once we made landfall on Antarctica, some of us summited a nearby mountain. From the summit, one of my climbing buddies pulled out an Iridium satellite phone and called his wife, "Guess where I am," he said. I was impressed that the phone worked and that he had the wherewithal to have brought it.

A few years later, I was crew on a Transpac sailboat race from Long Beach, California to Honolulu, Hawaii. About midway, roughly 1000 miles from both California and Hawaii, I pulled out my "sat phone" to call my wife. I looked around the 65' sailboat and noticed that two other crew members were doing the same thing, calling their families on their personal sat phones. What was phenomenally unique to me in Antarctica just a few years earlier now seemed commonplace.

According to figures from the U.S. Bureau of Economic Analysis and the Space Foundation, a nonprofit advocacy organization, the space business is about $400 billion per year, approximately 2% of our GDP. Put another way, $2 of every $100 in GDP represents activities in space business.

The space business has made a quantum shift from the clunky bureaucracy of state-run NASA to next level companies such as SpaceX, Virgin Galactic, Maxar, Blue Origin, and others.

**The major opportunities include:**

Spacecraft

Satellites

Networking

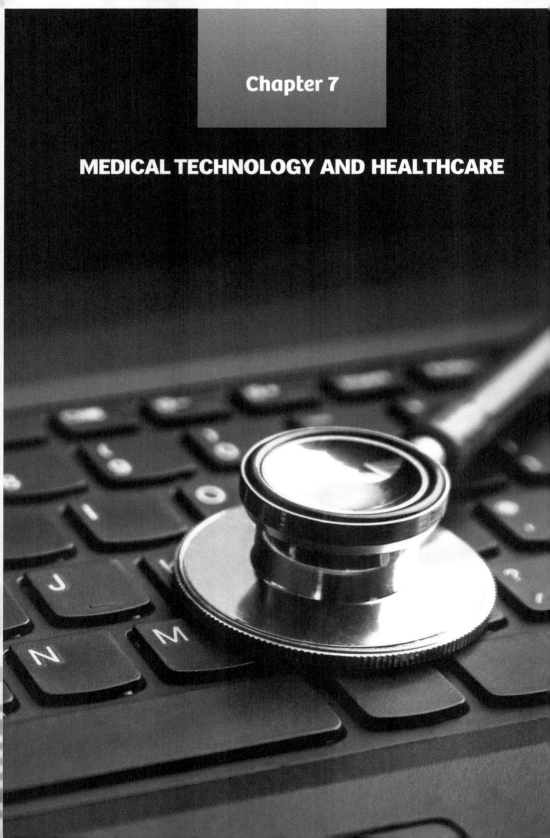

Chapter 7

MEDICAL TECHNOLOGY AND HEALTHCARE

U.S. healthcare spending reached $3.6 trillion in 2018: that's $11,172 per person and over 17 percent of U.S. GDP (2019, National Health Expenditure data-CMS).

Medicine is one of those areas in which technology can be especially disruptive and have a profound impact on people and society as a whole. For thousands of years, bloodletting and leeching were mainstream medical practices. The procedure was based on a flawed theory that humans possessed four humors or fluids (blood, phlegm, black bile and yellow bile), and that an imbalance in those humors resulted in disease. The manipulation of those fluids was thought to maintain good health or cure disease, and the way to manipulate blood was to withdraw it from the body, either though incisions made by a physician or through the application of leeches. Modern medicine began after the Industrial Revolution began in the 18th century. It wasn't until the later part of the 19th century that bloodletting and leeching ended.

Prior to the use of antibiotics, the average lifespan worldwide was 47 years. After 1928, when antibiotics became available through the invention of penicillin, that number rose to 78 (US National Library of Medicine National Institutes of Health). My paternal grandfather died in his early fifties, before the advent of penicillin, due to an infection that could have been curable by the antibiotics we have today. In the 1960s, my mother died in her mid-30s from breast cancer. Today, with modern treatments, over 27% more people survive breast cancer than just 25 years ago according to the American Cancer Society. As an industrial society and beneficiaries of the information age, we have accelerated our medical knowledge and abilities to treat illness, creating both longer lifespans and a better quality of life for many. Besides antibiotics, the big technology advances came in the following categories: vaccines, surgical anesthetics, antisepsis (anti-infection) techniques, clean water, improved sanitation, antiviral medication, improved heart surgery and cardiac care. Other breakthroughs occurred in randomized controlled trials, advanced imaging, advancements in childbirth, and organ transplants.

## Pharmaceuticals or "Pharma"

Drugs as a business, known as the pharmaceutical industry, has skyrocketed. According to Newsweek, large pharmaceutical companies are more profitable than large companies in most other industries (McCall, 2020). The Journal of the

American Medical Association (JAMA) investigated pharma companies that develop, manufacture, market and sell drugs, finding that between 2000 and 2018, 35 big drug companies received a combined revenue of $11.5 trillion, with a gross profit of $8.6 trillion. In a study that compared the profits of 35 large pharmaceutical companies with 357 large non-pharmaceutical companies from 2000 to 2018, the median net income of revenue was much larger for pharma companies than non-pharmaceutical companies (13.8% vs 7.7%).

## Medical Devices

Medical devices such as AFib pacemakers, advanced prosthetic limbs, high-tech knee and hip replacements, spinal devices, brain implants, automatic insulin dosing devices, and various cardiac catheters are just some of the "hard tech" that is available today that could significantly increase human lifespan and quality of life. Innovation continues, as thousands of studies for medical devices are submitted to the FDA for approval each year.

## Targeted and Controlled-Release Cancer Therapy

According to the National Cancer Institute, targeted drug therapy is a treatment protocol that utilizes drugs to identify and attack specific types of cancer cells. The goal of targeted, controlled-release cancer therapy is pharmaceutical delivery systems capable of providing targeted drug therapy to specific cells for a specific time period with little impact on healthy cells and without undue side effects. I liken this to using a rifle versus a shot gun.

Traditionally, surgery and radiotherapy have been the primary treatments for cancer, with anti-cancer drugs being used largely for metastatic cancers. Although chemotherapy has been successfully used for inhibiting cell growth for a long time, the side effects of chemotherapy have forced the drug industry to look for alternatives. The challenge has been to deliver the treatment to the area the disease has presented. While up until now, most drugs administered for this purpose have been infused, the emphasis has now changed to developing "needle-free" systems in an attempt to get away from injecting chemotherapy drugs. Some targeted therapies block the action of enzymes and proteins involved in the growth and spread of cancer cells; others help the immune system itself kill cancer cells. Targeted drug therapy may have fewer side effects than other types of cancer treatment, especially chemotherapy.

Cancer is the second leading cause of death in the US and is expected to pass cardiovascular diseases as the primary cause of death in the near future. Estimates by the National Institutes of Health (NIH) claim money spent on cancer in the US is expected to go above $150 billion by the end of 2020 (Targeted Cancer Therapies Fact Sheet. (n.d.). Retrieved from https://www.cancer.gov/about-cancer/treatment/types/targeted-therapies/targeted-therapies-fact-sheet). Smaller biopharma companies are involved in the development of new cancer drug delivery methods. Many of these companies are financed by venture capital with the goal of being acquired by one of the larger pharmaceutical companies for nice profit, often more than 10 times the original investment.

**Stem Cell Treatment**

According to the Mayo Clinic, "[s]tem cells are the body's raw materials — cells from which all other cells with specialized functions are generated. Under the right conditions in the body or a laboratory, stem cells divide to form more cells called daughter cells." (Mayo Clinic Staff, 2019)

The idea behind stem cell treatment is to generate healthy cells to replace diseased cells; the practice is also known as regenerative medicine. Stem cells are driven to become cells that can be used to repair diseased or damaged tissue. Stem cells can be used to treat those with spinal cord injuries, diabetes, Parkinson's disease, Alzheimer's disease, heart disease, stroke, burns, and cancer and may have the potential to become new tissue for use in transplants. Stem cells come from several sources, including embryos and adult stem cells. Adult stem cells have successfully been transformed into cells that can replace those damaged by chemotherapy or disease, or even help the recipient's own immune system fight some types of cancer and blood-related diseases such as leukemia, lymphoma, or myeloma.  By altering the genes in the adult cells, the cells can be reprogrammed to act similarly to embryonic stem cells. This technique may allow the use of reprogrammed cells instead of embryonic stem cells and prevent immune system rejection of the new stem cells.

**Medical Software**

Medical software is a broad category that includes patient-related health care information technology as well as software designed to manage the complicated data and service needs of hospitals, medical practices, and research. Medical software can be found in most areas of healthcare and the medical industry as a whole, including pharmacies, labs, and clinical facilities. Bigger medical practices and medical institutional facilities use multi-tool software solutions to aggregate patient information. These tools include business-oriented resources to maintain record keeping as well as facilitate increased engagement between patients and their providers through messaging, appointment setup and reminders, and even patient education. We've seen during the COVID-19 pandemic that telemedicine can be a key component of ongoing medical care. Increasing these touchpoints can help providers expand their businesses as they work toward improving care, and technology is poised to be a key component of this shift, offering many future expansion opportunities to create a better interface between the patient and their practitioners, labs, and medical records.

The industry is not yet mature, and there is lots of remaining opportunity. A 2018 Stanford University survey found that while 63 percent of physicians feel that current electronic health records (EHR) contribute to better patient care, 71 percent feel that EHRs contribute to physician burnout and 59 percent feel that EHRs "need a complete overhaul" to decrease physician stress. Medical business software can positively impact the medical industry in the following ways:

Reduction in clerical errors

Increase in revenue

Decrease in insurance denials

Greater control over revenue cycle

Behavioral health software is distinct medical software for mental and behavioral health professionals. These practitioners often work in hospitals, private offices, and outpatient clinics. The needs of this group require a custom software solution. Various software tools allow medical professionals to connect with their patients through applications on mobile devices to track patients' moods, condition, and progress.

With the expansion of software into wearable devices (wearables), medical software has gone past the healthcare sector and into lifestyle. Wearables measure and store vital signs and health data points, and can be used with other data to assist in diagnosis and patient care and engagement. From cardiac monitors to blood sugar meters and insulin pumps, the Fitbit to the Apple Watch, more and more people are walking and running through each day, gathering data and sometimes using that data to shift their behaviors toward better health.

Stem cell treatment, medical devices, and new pharmaceuticals are revolutionizing modern medicine. These technologies have been very profitable, and, in many ways, they are in their infancy. Medical software and the incorporation of technology into our daily lives is a related area that shows great potential. There's a long way to go with a myriad of applications for these medical technologies, and research continues to advance knowledge and applications.

The opportunities in the fields of medical technology are:

Pharmaceuticals

Medical Devices

Drug Delivery Systems

Stem Cell Technology

Medical Software

**Chapter 8**

**RENEWABLE ENERGY**

U.S. energy expenditures per GDP reached 5.8% in 2017. Average U.S. prices for petroleum and natural gas increased by 14% and 13%, and electricity prices increased by 2% year over year. Total U.S. energy consumption increased by about 1% in 2017 (U.S. Energy Information Administration).

In elementary school, we learned about the perpetual motion machine, a device that would create its own energy through the production of creating the energy.

"That dog will do anything for a biscuit."

The idea falls on its face based on Newtonian physics: energy has to come from somewhere. It can come from muscle power such as rowing or pedaling, it can come from wind, water flow, tidal change, or from our nearest star, the sun. It can also come from chemical or nuclear reactions such as those created by diesel or gas generators, burning coal, and nuclear power plants. We know that there is a direct link between fossil fuels and climate change. This reality is driving a shift to renewable energy sources such as hydropower, solar, biomass, geothermal, and wind. Let's look at some different types of energy and discuss some drawbacks and the potential of each.

## Nuclear Power

Present-day nuclear power plants are fission reactors, creating power by splitting atoms. These plants are considered dirty and dangerous based on their risk of radiation exposure resulting from operator error, mechanical failure, or natural disaster such as what occurred in 2011 at the Fukushima Daiichi Nuclear Power Plant in Japan. The failure was a reaction to a tsunami that damaged the reactor causing a serious accident with death, injury, and massive property damage instigating evacuation of the area. A similar incident occurred in 1986 in the former Soviet Union, now Ukraine. The Chernobyl disaster was caused by an accident that resulted in a meltdown and abandonment of the facility.

Nuclear fusion is different from the fission reactors in Chernobyl and Fukushima. Fusion is a phenomenon in which two or more particles are combined to form one or more different atomic nuclei and subatomic particles. The difference in mass between the reactants and products arises from either the release or absorption of resultant energy and is due to the difference in the binding energy between the atomic matter before and after the reaction. Fusion is the process that powers stars. It is interesting to note that there is no radioactive waste byproduct of fusion. Scientists in China have created a fusion reactor. In 2018, the Chinese fusion reactor became the first in the world to reach 100 million degrees Celsius.

### Wind as a Resource

Wind as a power source may have been recognized as early as 1200 A.D. by the Chinese, and later by the Persians in 900 A.D. The technology came to the U.S. in 1854, brought by Daniel Halladay. Early windmills powered grain mills and water pumps. The first windmills that generated electricity were built in 1888 in Cleveland, Ohio by Charles F. Brush. In the 1970s the U.S., and especially California, took the lead in windmill tech. However by 2000, Europe surpassed the U.S. in wind energy manufacturing (Third Planet Windpower).

Wind is a clean and renewable energy source. Wind energy is actually a form of solar energy: winds are caused by the heating of the atmosphere by the sun, the rotation of the Earth, and the Earth's surface variations. As the sun shines and the wind blows, the energy created can be sent across the power grid. The nation's wind supply is unlimited. In fact, wind is now the largest source of renewable power in the United States.

Using wind to produce energy has less adverse impact than many other energy sources. The creation of wind energy doesn't emit particulate matter, nitrogen oxides, or sulfur dioxide—substances that can cause health problems and economic damage. Wind turbines don't produce atmospheric emissions that cause acid rain, smog, or greenhouse gases. Wind turbines do not release emissions that can pollute the air or water and they do not require water for cooling (Office of Energy Efficiency & Renewable Energy). Wind turbines can be built on existing farms or ranches; because the wind turbines use only a fraction of the land, farmers and ranchers can continue to work the land. This benefits the rural economies where many wind farms are found.

One of the lowest-priced energy sources available, wind power costs less than 2 cents per kilowatt-hour after current production tax credits. Electricity from wind farms is sold at a fixed price over a long period of time and since the energy is basically free, wind energy lessens the price variability that fuel costs add to traditional sources of energy. Wind energy also creates jobs. From the Office of Energy Efficiency and Renewable Energy website: "The U.S. wind sector employs more than 100,000 workers, and wind turbine technician is one of the fastest growing American jobs. According to the Wind Vision Report, wind has the potential to support more than 600,000 jobs in manufacturing, installation, maintenance, and supporting services by 2050." Wind projects account for more than $10 billion in the U.S. economy. The United States has a skilled workforce that can compete globally in the clean energy economy.

There are, however, some problems with large-scale implementation of wind energy. Wind power has to be economically viable. There has to be enough wind to make the turbine work and the wind created electrical energy has to be able to compete with the existing electrical sources, which are often less expensive. The infrastructure has to be improved. New transmission lines need to be built to bring the power from the wind turbines to where it's needed. Another challenge is land use: land could potentially be better used for other things and some people think turbines are ugly and noisy. And finally, wildlife, namely birds and bats, gets killed by the blades.

## Solar

In 1884 Charles Fritts used 1% efficient selenium cells on a New York roof. After 1884, solar power fell out of favor because of cheap petroleum and coal. Though solar energy technology has been around since the 1800s, solar power has recently become very popular. The energy crisis and fuel embargo of the early 1970s caused concern over fossil fuel supplies and pricing causing a reinvigoration of solar technologies. In 2010, the International Energy Agency estimated that 27% of global energy could be created by solar power. According to the NASA technology transfer program, today's solar technology efficiency has increased to 30-40% (Technology.nasa.gov, High Efficiency Solar Cell).

**Bringing it Home**

How practical are renewable energy sources and systems for individuals? I challenged myself to figure out how to save money and use less energy at the same time. I had solar panels installed on the roof of my home, and I purchased a Tesla Powerwall (battery) to store energy. If the power goes out at our house, we can function almost indefinitely, albeit at a lower draw rate. With only one Tesla Powerwall, we can run our stove, refrigerator, lights, and basic outlets. It's also possible to use the batteries to meter power into the house during expensive peak times to mitigate cost and utilize less of the common grid power.

I got the idea from my sailboat races from San Francisco to Hawaii. We had to figure out what our electrical draw would be to run our computers, radios, and lights. I created a spreadsheet to compare battery storage with daily electronics usage and came up with an energy plan. In our first two-week race from San Francisco to Hawaii, diesel fuel was a limiting factor, as we had to run a small diesel engine a couple of hours a day to keep the batteries topped. For our second race a couple of years later, I installed solar panels on the back of the boat to help charge the batteries to save fuel and weight. The goal was to give us a better chance of winning. What I did not know from the start was that we could make the trip without using the motor at all: we could use the solar panels to keep the batteries minimally charged all the way to Hawaii. The net effect was that we only ran the engine a little, mostly on overcast days when we did not have direct sun to allow the panels to efficiently charge the batteries.

This sailboat race provided me with the inspiration to power my house the same way. I used solar panels to augment my power and provide storage in case of a power failure or worse, a long-term power shut down because of an event such as an earthquake or other disaster. In fact, one of my main motivations to get an electric motorcycle was to take advantage of my home's solar panels. Instead of getting an electric car that requires a significant amount of energy to charge, I got the bright idea to purchase an electric motorcycle. Because it's lightweight and streamlined, the motorcycle is very efficient and quite fast. The bike goes up to 200 miles on a single charge, and usually takes less than an hour to charge. It has similar performance to a top end Tesla automobile at a fraction of the cost. And I can charge it basically for free from my home's solar panels.

My thinking wasn't particularly unique. Many households and communities get off of the grid. Hawaii, however, is the only state that allows its citizens to get totally off the grid. Many rural areas and businesses take advantage of various alternative power technologies to mitigate their power use and save money.

## Batteries

Generating power is only part of the equation: it's critical to be able to store the energy for later use. Batteries are integral to many new technologies, from modern self-powered suitcases to spacecraft. They can be charged by outside sources such as solar panels or the electrical grid, or by a chemical reaction.

Elon Musk's Tesla is an obvious beneficiary of revenue from batteries. The battery pack for a Tesla is made of 16 lithium-ion battery modules, each containing 516 battery cells for a total of 8,256 battery cells. Multiply that by 500,000 electric cars by 2020 to reach 4.1 billion cells. With the demand for batteries growing, Tesla plans to triple its capacity and is planning to build as many as three more factories in the years to come. Panasonic is Tesla's partner in the battery business. The Japanese electronics company occupies a significant portion of Tesla's Nevada battery factory. The more batteries Tesla sells, the more money Panasonic makes.

China is racing to create battery factories of its own. Chinese electric car company BYD is another big player in batteries. At the time of writing, Warren Buffett's Berkshire Hathaway owns a 16.5% stake in BYD and is the company's second-largest shareholder, according to data from S&P Global Market Intelligence. BYD is a huge corporation, involving more than just electric cars. Their home page says they are "...dedicated to providing zero-emission energy solutions."

A leader in specialty chemicals, Albemarle is heavily leveraged in the growing popularity of lithium-ion batteries. These batteries are profitable products for Albemarle, earning a net profit margin of 27% — after tax. According to Mining Global, Albemarle is now the world's top producer of lithium. What's more, in January 2020, it secured a deal with Chile's Development Agency allowing it to grow its lithium mining operations in that country, with the result that Albemarle's production of "battery-grade lithium carbonate" is expected to more than triple to 80,000 tons per year. Analysts who follow Albemarle are telling investors the company will only grow its earnings about 11.4% annually over the next five years. With supply from its mining operations expanding, though, and demand from companies like Panasonic and Tesla growing, I think that estimate might prove conservative.

The U.S. Chamber of Commerce's Global Energy Institute makes the following conclusions based on the (IEA) World Energy Outlook 2018 (WEO2018) energy forecast (https://www.iea.org/reports/world-energy-outlook-2018):

**Demand Growth**

Energy demand from 2017 to 2040 is expected to grow by 27% globally. Expectations are that developed country demand is on course to shrink their demand by 4% while developing country needs will increase from 64 to 70%.

**Fossil Fuels**

Fossil fuel demands are forecast to grow by about 16% by 2040, with natural gas accounting for 43% of that, petroleum at 10%, and coal at 2% according to the 2018 study. Although the share of energy demand met by fossil fuels decreases over time, hydrocarbons still account for 74% of total global demand in 2040 compared to 81% in 2017.

**Renewables**

Demand for renewable energy is expected to increase about 81% by 2040. Coal will likely be the big loser. Energy demand met by renewable technologies should climb to roughly 20% by 2040. Hydroelectric power will remain the largest single source of renewable energy, accounting for 50% of renewable electricity output in 2040. By 2040, solar energy will supply nearly nine times as much energy as in 2017. Wind energy will grow to 400% of its 2017 level. Bioenergy (energy derived from plant and animal materials, also referred to as "biomass") will remain the most widely used renewable energy source. It accounted for approximately 70% of renewable energy in 2017 and is expected to account for about 50% in 2040. These renewable technologies will likely be used in developing countries, accounting for approximately 69% of the total increase in renewable energy consumption globally.

**Nuclear**

Nuclear power is likely to grow its overall share of demand more than 40% by 2040. Most of that increase is anticipated to come from Asia and the Middle East, at least partially due to Japan's return to nuclear-based power generation after its nuclear shut down following the Fukushima Daiichi accident in 2011.

**$CO_2$ Emissions**

Based on the emissions pledges made by various countries such as China and India, it's not clear how much $CO_2$ emissions will actually be in a few decades. However, China has said it will peak its $CO_2$ emissions by around 2030, though projections indicate China's emissions will continue to climb through 2040, increasing by nearly one-third from the 2012 level. Over the same period, emissions from India will likely at least double while emissions in Africa (89%), Southeast Asia (91%), and the Middle East (82%) will also increase. If it is assumed that developed countries such as the United States and Europe achieve their pledges, the increases in emissions forecast from less developed countries will more than offset, by a large margin, reductions from developed countries, thereby worsening the situation.

The future of renewable energy is potentially limitless. This area will likely be one of the most powerful areas of investing in a new climate. Nations participating in the Paris Agreement share a goal of limiting global warming to less than 2° Celsius above pre-industrial levels. The only way to achieve this is to reduce carbon in the atmosphere to mitigate the greenhouse effect. This reduction will require a monumental effort and an unprecedented cost to implement alternative and renewable energy sources. Opportunities include the following areas:

Solar

Wind

Electric

Hydraulic Energy

Nuclear Power

# Chapter 9

# REAL ESTATE AND CONSTRUCTION

In 2018, real estate construction contributed to more than $1 trillion to the nation's economy. That equals 6.2% of U.S. GDP (Amadeo, 2020).

Generally speaking, real estate is defined as property consisting of land and buildings. According to www.fastexpert.com, "the term "real estate" was first recorded in the 1660s and holds the oldest English sense of the word. ... "real" is derived from the Latin meaning of existing, actual or genuine, and "estate" refers to the land." Geography is also important: where is the property? The old joke is that real estate has three important aspects: location, location, and location. This is perhaps more true now than ever. As the world heats up and sea level rises, where do we go? And where do we invest? Is Northern Europe a better bet than sub-Saharan Africa? Is Canada a better place to invest than Mexico? Higher latitudes will likely fare better than equatorial regions. In general, places that are already successful will generally do better than those places that are struggling.

For the purposes of this book, we will take a broader view of real estate. We will include geography such as continents, hemispheres, countries, states and territories, as well as localities. We've discussed various climatic disruptions, such as volcanoes, droughts, and superstorms: where will disruptions occur and how should we deal with them? We have discussed areas that will be most impacted by climate change such as sub-Saharan Africa, low-lying island nations, and coastal locations subject to the adverse impacts of rising sea level. There are also other factors such as unusual weather, such as the California 2020 rare "dry lightning strikes" in the summer months triggering uncontrollable fires in drought-ravaged old growth forests. While California has suffered lightening driven fires in the past, 2020's lightning fires are an anomaly according to Rick Carhart of Cal Fire. Scott Rowe, a meteorologist with the National Weather Service, claimed that the recent surge of lightning-caused California fires has been one of the most noteworthy in a decade.

In 2018, the violent volcanic explosive eruption of Hawaii's Kilauea sent ash 30,000 feet into the atmosphere, which caused not only disruption to air traffic but caused a virtual shutdown of the island of Hawaii, the state's largest island. The damage from the volcano destroyed dozens of homes and made the air quality so bad, people had to stay indoors and tourism came to a halt. While not directly associated with climate change, the eruption highlights the impact of local hazards. The experiences of those living near erupting volcanos, in Tornado

Alley, and areas where hurricanes regularly destroy homes and businesses can lead us to carefully consider where we choose to live and invest.

I flew missions for the US Geological Survey (USGS) in the early 1980s. We spent a lot of time near Mammoth Lakes in California's Owens Valley, just east of the Sierra Mountain Range. The team I was part of also included scientists studying the possibility of a volcanic eruption near Mammoth Lakes. At that time, there was a concern that an eruption could occur that could exceed the Mount St. Helens eruption in both magnitude and disruption. Airborne volcanic ash can be carried hundreds of miles downwind, and though the amount and size of falling ash decreases with distance from the eruption site, even a light dispersal of volcanic ash can close roads and seriously disrupt communications and utilities for weeks or months after an eruption (USGS, Survey Fact Sheet 073-97, V 1.1). Based on the area's prevailing winds, an eruption near Mammoth Lakes could negatively impact Southern California for weeks or months.

The technology we have today allows us to mine data to forecast future weather patterns, predict seismic and volcanic activity, and accurately predict sea level rise, delivering better insight than has ever been available into the future of real estate opportunities and threats.

With a focus on California, I spoke with Redwood City-based BKF Civil Engineers, Roland Haga and Yousra Tilden. BKF has been a California civil engineering firm since 1915, specializing in infrastructure. While it's common knowledge that earthquakes are a big threat to California's population, Haga and Tilden highlighted flooding, hurricanes, superstorms, seismic events, and extreme temperatures as the biggest and most likely disruptions to potentially impact real estate in California. Flooding is becoming a bigger potential problem in California. The combination of superstorms and high tides, happening simultaneously, could have dramatic impacts on the population. In coastal areas in California, the law requires structures be built at least 13 feet above sea level unless the land is protected by a levy. According to a 2017 study by a working group of the California Ocean Protection Council Science Advisory Team (OPC-SAT), California's sea level could rise from current levels to a few feet to 10 feet or more by the year 2100.

The three Inyo Craters, part of the Mono-Inyo Craters volcanic chain, stretch northward across the floor of Long Valley Caldera, a large volcanic depression in eastern California. During the past 1,000 years, there have been at least 12 volcanic eruptions along the chain, including those that formed the Inyo Craters and South Deadman Creek Dome (seen here just beyond the farthest Crater). (USGS)

Haga pointed out that, while it's possible to build structures such as homes to very high standards so they might survive moderate or worse earthquakes, the cost of materials makes it prohibitive to do so. Because of this, it's generally accepted in the construction industry to build to code and perhaps a bit more, but not to build to withstand the worse possible scenario. Therefore, the real window into the real estate construction supply market is current and proposed government regulations. Should the government decide to change the rules in building residential or commercial property, whether the decisions are coming from a local, state, or federal perspective, it could be very beneficial to keep up with those rules. The zoning and government building criteria are pivotal in deciding what procedures, products and building materials will be required to comply with those new or proposed rules. Should a new type of construction material be required, perhaps we should invest in the companies that make those materials.

The short-term opportunities lie in forward-thinking businesses that are producing products and services that capitalize on modern construction techniques and supplies. The Constructor, a reference website for civil engineers, provides a comprehensive list of sustainable building materials that can be used in almost any part of construction, from foundation to roof (www.theconstructor.org). Some of these materials are high-tech, while others are low-tech. They include materials that have been used for millennia, such as stone, sand, earth, wood, cork, and thatch, and more modern engineered products like non-VOC paints, composite shingles, engineered wood products, recycled steel panels, polystyrene, and structural insulated panels or SIPs. Depending upon the specific properties of the material, sustainable options offer excellent insulation, sound-proofing, strength and durability while often impacting the environment less than non-sustainable options.

Once you've chosen the right building materials for your project, where do you start construction? What is the importance of location in real estate? Increasingly, location —on a macro scale— should be a critical part of the decision process when investing in real estate. We have identified a variety of possible disruptions to geographies around the world: temperature changes, flooding, severe storms, chronic fires, and volcanic eruptions. Dealing with these disruptions alerts us to avoid owning property in high risk areas. We should also be open to opportunities in areas that have moderate climates and less risk of adverse impacts. The higher latitudes away from low lying seashores look appealing, such as areas in the Northern U.S. (including Alaska), Canada, Northern Europe, and New Zealand.

**What's the future of real estate, and where are the opportunities?**

Land

Property

Construction materials and supplies

**Chapter 10**

# COMPUTERS AND INFORMATION TECHNOLOGY

The digital economy accounted for 6.9% of the U.S. GDP, or $1.35 trillion, in 2017, according to the Bureau of Economic Analysis. Technological advances associated with modern and next generation computing will very likely have a profound impact on the world. These impacts include but are not limited to brain mapping, targeting pharmaceuticals, transportation efficiencies, financial optimization, and climate modeling.

In 1965 Gordon E. Moore, the founder of Intel, postulated that the transistors packed in a given space doubles every two years. Known as Moore's Law, the essence of the concept is that computers get twice as powerful about every two years. While these days, with silicon chips, capacity doubles approximately every 18 months, Moore's Law has been more or less true for the last 50 years. There's an argument lately that this phenomenon has slowed down a bit. In fact, in 2005, Moore himself predicted the so-called law would eventually hit a wall.

But not so fast. Since early computers were designed and implemented, such as Alan Turing's famous Nazi code breaking computer in 1936, modern computers have been improving constantly. The very first supercomputer was built in around 1946 by U.S. physicist John Atanasoff. Atanasoff's computer boasted 18,000 vacuum tubes versus the 2000 tubes of its predecessor ENIAC (Electrical Numerical Integrator and Calculator). Effectively, serious computers began in the 1950s with vacuum tubes, the 1960s brought transistors, the 1970s integrated circuits, and silicon chips since the 1970s.

In my opinion, whenever we think technology has stagnated, we should watch out for what is coming next. In this case, quantum computing. Quantum computing and merging classical computers with quantum computers will revolutionize computers in ways we have not yet imagined.

**Quantum Computing**

Quantum technology is not new. The technology was pretty well identified and coalesced in the 1920s. A pivotal demonstration that really started the whole thought experiment of quantum physics was the Double-Slit Experiment, demonstrated in 1801 by Thomas Young. The double slit experiment goes like this: if you shoot light particles through two slits, one would expect that the effect would be two strips of light appear on the wall behind the two-slit device. But that's not what happens. Instead, the particles form an interference pattern that acts more like waves than particles, creating many strips of light. The experiment shows that very small particles can act strangely, they can be in two places and/or in two forms at the same time: in this case, particles and waves. To make the experiment even more difficult to explain, when the double slit experiment is observed, the waves return to being particles and the interference pattern does not appear. Looking at the particles changes the outcome. The simple act of observation actually changes the outcome. Physicist Richard Feynman said that the result of the double slit experiment was "a phenomenon which is impossible to explain in any classical way, and which is the heart of quantum mechanics."

And then there's quantum entanglement, which is when two particles that are not together — even separated by great distances — act in unison. When one particle acts a certain way, the other entangled particle acts the opposite way. Einstein originally did not like this concept, sarcastically calling it "spooky action at a distance." Later, the phenomenon of entanglement was empirically proven, contrary to Einstein's view.

Why am I mentioning these complicated concepts? These two fundamental concepts allow quantum computing to work. The basic difference between classical computers and quantum computers is that while classical computers use bits, or 1s and 0s, to make calculations, quantum computers use 1s, 0s, 1s and 0s, and 0s and 1s all at the same time. The quantum bits are called qubits. In comparison to the classical 1 and 0 bit, each qubit doubles the states of information that can be stored: two qubits can store four states, and three can store eight. Fifty entangled qubits can store the equivalent of 1,125 quadrillion classical bits.They can do calculations at much faster rates and deal with much more data than classical computers. Google has announced that its quantum computer is 100 million times faster than any classical computer in its lab. Google has a quantum computer division, as do IBM and a company called D-Wave.

There are two ways of using quantum computers: you could buy the computer and keep it super-cold or access the computer through the cloud. To avoid "noise" that negatively impacts the quantum computation process, the computer needs to stay at just above zero kelvin, which is about minus 460° Fahrenheit, colder than outer space. Therefore, it's arguably more practical to access the quantum computer remotely via the cloud through an internet connection than buying and maintaining your own. IBM offers a free service called the IBM Quantum Experience. This interface allows schools, scientists, and anyone to access their computer via the internet. The access is relatively easy, primarily requiring the pointing and dragging of icons. The output delivers the answer to formulas based on probabilities. While certainly at its practical or useful infancy, the system is online now, and it works.

Many large companies have made significant investments in quantum computing technology. Besides IBM and Google, Intel has been a big early adopter. There are half a trillion dollars in venture capital being invested to explore the opportunities that quantum computing may offer. That's a lot of serious people and serious money chasing this new technology.

D-Wave is a smaller Canadian company that has sold multimillion dollar quantum computers to financial and other sectors to develop leading-edge optimization technologies. The D-Wave process is controversial in that it could be considered a hybrid classical-quantum computer that may create a result only marginally better than advanced standalone classical computers. But from a business perspective, it may only take a small improvement over classical computer technology to effect a big difference in efficiencies and profit.

While this technology was arguably conceptually invented in the beginning of the last century, its implementation is only now seriously beginning to take hold. In reference to the potential of quantum computation, I compare where we are today to a 1982 Commodore 64. Another example is early space shuttle computers, which had less computing power than a smart phone has today. In my opinion, we are on the brink of a computer revolution that could cure cancer in specific individuals, forecast dangerous weather events sooner and more accurately, and optimize a myriad of business and government applications. But while the possibilities of quantum computing are infinite, many computations with smaller data sets can be performed just as well on classical computers. The most promising use of the quantum computation process is accelerating the study and use of quantum science itself, leading to even more revolutionary technologies and applications. So, what can quantum computers do now? Let's look at some examples, starting with D-Wave's offerings.

**Machine Learning and Computer Science**

Finding errors in complicated statistical data

Increasing information flow by compressing data

Using patterns and images to enable computer recognition of patterns and images

Teaching synthetic neural networks to reach higher levels of efficiencies

Building and verifying new and existing software

Fixing circuit problems

**Financial Modeling**

Detecting market instabilities

Developing trading strategies

Optimizing trading trajectories

Optimizing asset pricing and hedging

Optimizing portfolios

**Security and Mission Planning**

Detecting computer viruses and network intrusion

Scheduling resources and optimal paths

Determining set membership

Analyzing graph properties

Factoring integers

**Healthcare and Medicine**

Detecting insurance fraud

Generating targeted cancer drug therapies

Optimizing radiotherapy treatments

Creating protein models

IBM put the first quantum computer in the cloud in 2016, moving from a few users to thousands of users. Their IBM Q Network now brings together more than 100 Fortune 500 companies, academic institutions, labs, and startups. Delta Airlines (DAL) has partnered with IBM to use biometrics for its international operations. DAL will access the IBM Q Network through NC State University, giving them access to a 53-qubit quantum computer, which currently has the most qubits of a universal quantum computer available. The term "universal quantum computer" identifies the technology as a true quantum computer, not a hybrid. IBM is also working with Daimler AG, the parent company of Mercedes-Benz. Electric cars have an intrinsic weakness: batteries. IBM and Daimler used quantum computing to model the dipole moment of lithium molecules to create a better next-generation battery technology. With partners like DAL and Daimler, IBM has become a serious quantum computing player.

The quest for "quantum supremacy" has had some setbacks. Strangely, quantum technology has spurred a boom in quasi-classical computing technology, making classical computers more powerful, effective, and efficient. Quantum technology works better with some algorithms than others. And it's not all about the hardware: the software is a big player for quantum computers as well to interface with the process to realize an advantage over classical computers. Software for quantum computing is a big player and has a long way to go. In 1981, Richard Feynman said, "Nature isn't classical, dammit, and if you want to make a simulation of nature, you'd better make it quantum mechanical."

## Optimization

Mathematical optimization is a collection of mathematical principals that can be used to solve quantitative problems in many disciplines. Long-distance offshore ocean sailboat races are won and lost using this type of optimization software. I learned this first-hand during two races to Hawaii, in 2012 and 2014. After getting the boat ready and the crew trained, the real work began: setting up a computer system and software to be competitive.

The race takes from about 10-17 days in a sailboat of my type. We needed to constantly get updated wind data into the computer to work with the polars to optimize our route. We have to analyze how well our boat sails at different angles to the wind: these angles are called polars. In mathematics, polars are a two-dimensional coordinate system where each point on a plane is determined by a distance from a reference point and an angle from a reference direction. For example, in 10 knots of wind, with the wind coming from a 45% angle from the front of the boat, my boat would perform at about 5 knots. With a 20-knot wind coming from behind the boat, the boat would travel at about 8 knots. At each intermediate angle and wind speed the boat would perform slightly differently. The software knows how my boat performs at different wind angles and wind speeds: to provide optimized headings for us to sail, I have to get the current and forecasted winds into the computer. I download wind charts from the national weather service via satellite. I hit the optimize button and voila: the computer provides a spreadsheet of headings to sail each 20-minute period for about two weeks. This is updated every six hours as new weather reports become available.

Optimization can also be instrumental in finance, and there are lots of financial tech (fintech) tools available for investors and advisors to help make plans and decisions regarding specific investments. As we age as investors, our approach to investing and our risk tolerance evolve. So, in our early 20s we might have 90% of our money in the stock market and 10% in bonds and cash. We generally get a good return, but we suffer high volatility. In our 40s and 50s we may have about half of our money in stocks and half in bonds and cash, since we can tolerate less volatility. Once we get into our late 70s, we generally do not want to suffer much volatility at all since we are probably retired and living off the proceeds of our investments. This leads to an 80% cash and bond position, with perhaps 20% in stocks. But what I just explained are estimates, based on anecdotal assumptions.

Financial planning professionals generally run computer-based simulations and optimization programs to fine-tune asset allocations. For example, a 46-year-old married woman working in a profession with a mandatory retirement age, with 3 children, a non-working spouse and a conservative investment philosophy might come up with a specifically optimized portfolio allocation designed just for her. That portfolio could have 54% of mixed equities, 32% in corporate and municipal bonds, 7% in certificates of deposit, and 7% in cash. It's possible, and in fact common practice, to provide investment portfolios to clients that are computer optimized.

These are specific applications of optimization in the worlds of navigation and finance. Consider other categories where optimization could be beneficial: business, biological applications, physics, and engineering. We can imagine that, as technology continues to become more advanced and as quantum computing capability becomes more accessible, optimization should provide efficiencies we as a society have only dreamed of. These advances could help us mitigate air pollution, stretch out energy potentials, save natural resources, improve labor costs and efficiencies, improve transportation efficiencies, and on and on.

**What's the future of computing, where are the opportunities?**

Hardware

Software

SaaS (Software as a Service)

Cloud Computing

Cyber Security

# Chapter 11

## ARTIFICIAL INTELLIGENCE AND ROBOTICS

The global artificial intelligence market size was estimated at $39.9 billion in 2019 and is expected to reach $62.3 billion in 2020.

Back to my boat race to Hawaii. Usually thought of as an old-fashioned and low-tech endeavor, the reality of offshore sailboat racing is fraught with high-tech navigation subtleties that make or break the outcome. I needed to set up my thirty-foot sailboat to communicate with other boats, talk to the mainland, and download and analyze data. In our earlier discussion of using computers to optimize various processes, I described the software I used to analyze wind charts, weather service data, and my boat's performance to optimize course headings. In the old days, the captain of a boat would plot a course from point A to point B, consider the weather by looking at the cloud patterns, the winds aloft as seen from the deck, and take a good look at the sea state. Then the captain, or perhaps the navigator, would use that information to come up with a game plan that might include trying to avoid storms, looking for favorable winds or maybe just "going for it" by sailing a straight line to the destination, sometimes with disastrous results. While the old ways of celestial navigation, dead reckoning, and shooting from the hip based on nautical experience and acumen are romantic, those practices do not generally prevail in modern-day sailboat racing. The human approach needs to be augmented. The augmentation comes from artificial intelligence.

The fathers of the field of artificial intelligence, Harvard PhD Marvin Minsky and Stanford professor John McCarthy, described artificial intelligence as any task performed by a program or a machine that, if a human carried out the same activity, we would say the human had to apply intelligence to accomplish the task. Artificial intelligence can be seen both as a tremendous opportunity and as a monumental disruption, akin to taking out some of the romance from sailboat racing. Commonly known as AI, artificial intelligence is a tool that can address topics such as climate change and weather prediction in a revolutionary way. Google claims that new AI models nearly instantaneously forecast weather. We're not talking about weather observations: this is prediction and forecasting almost instantaneously, seeing into the future. Through AI, we can know what the weather will do sooner, faster, and more accurately than ever before. Like quantum computing, artificial intelligence has its own technological momentum and is a big part of investing in the future.

Part of the benefit of AI is that it has no traditional human bias and it "thinks" fast and accurately. AI can allow people to pursue things outside of work and help take care of family all at the same time. How many of us check our emails on the beach or from vacation without bothering to place a standing out of office email message? Who would have imagined even a decade ago that we could work effectively remotely with email, texting, Webex, and Zoom? The COVID-19 pandemic has taught many of us what is possible, working and even living remotely. Most of us use AI every day, often all day. The iPhone and other smart phones ooze AI and people cannot put them down, ever. We walk with them, we run with them, we drive with them, we eat with them. Today's iPhones have more computing capability than the computers that led the manned Apollo 11 mission to the moon. To put it another way, the iPhone is 120,000,000 million times faster than the Apollo-era computers. According to ZME Science, "...you wouldn't be wrong saying an iPhone could be used to guide 120 million Apollo era spacecraft to the moon, all at the same time..." (https://www.zmescience. com/science/news-science/smartphone-power-compared-to-apollo-432/) We see AI in an iPhone when the screen lighting automatically increases when you lift up the phone and dims when you put it down, an accelerometer senses speed, route suggestions appear in maps. Apple's online help feature Siri reeks of AI. Text response suggestions generate automatically when we receive an incoming text or are in the process of creating a new text. Auto scheduling suggests open time slots in your schedule. Facial recognition, voice recognition — all further examples of AI in use today.

But what exactly is artificial intelligence? There is some subjectivity in that question to be sure, but one popular definition is designing computers to think and learn like a human would. Artificial intelligence has been around for a long time. According to Science in the News, a Harvard Publication, written by Rockwell Anyoha in 2017, the AI concept started in 1899 with the Tin Man in The Wizard of Oz, followed closely by a humanoid robot in the 1927 German film, Metropolis. Alan Turing, who developed the Turing Test to determine whether machines can think, suggested "...that if humans can solve problems and make decisions, why can't computers?"

What have we achieved since Dorothy returned to Kansas? Between 1957 and 1974, AI gained steam making information more accessible, quicker, and cheaper. In 1970, Minsky told Life Magazine that in "...from 3 to 8 years we will have a machine that will have a general intelligence of an average human being." Well, that didn't happen. By the early 1990s, funding for AI basically dried up. Ironically, after that, perhaps because of the efforts that failed, earlier attempts excited and motivated scientists to continue their work and achieve big strides. In 1997, an IBM computer called Deep Blue beat Grandmaster Gary Kasparov at chess. In the same year, speech recognition software became commercially available using Microsoft Windows. Also in the late 1990s, a robot called Kismet, the brainchild of MIT's Dr. Cynthia Breazeal, could recognize and simulate human emotions.

In the last 20 years we've come a long way: as predicted by Moore's Law, computation capabilities have exponentially increased in the last two decades. Breakthroughs have manifested themselves in computer science, mathematics, medicine, neuroscience, and physics. AI has become pervasive in our society: word recognition, voice recognition, Siri, Alexa, transportation apps, spam filters, plagiarism checkers, and mobile check deposit systems are real-time applications of AI that many of us use every day.

AI is also being used by investment advisors to give automated advice to their clients based on millions of data inputs, optimal pricing, and economies of trading costs. These robo-firms have revolutionized investing for many individuals and institutions. The outcome is better diversification, faster portfolio rebalancing, and lower costs. Younger, tech-savvy investors have flocked to this new technology based on its ease of use, transparency, and costs.

This list of potential AI uses doesn't even scratch the surface. Other applications could include "smart autopilots" for airliners allowing fewer required pilots necessary for flight and allowing for greater aviation safety. Autopilots for cars have been developed by Tesla that teach all Teslas (at least those with autopilots) to learn of road changes, obstacles, and other hazards that appear in real-time. If the lane line in the road disappears, perhaps due to road construction, and the Tesla autopilot's navigation system disengages, the Tesla network finds out and updates all other autopilot driven Teslas of the new road issue. Then, as other Tesla vehicles navigate that piece of road, they compensate for the lane line anomaly based on the information from the previous Tesla.

Human augmentation is making a person better, or perhaps fixing or improving someone with a physical challenge or injury. Some examples could be a creating a new limb that allows for a more complex neural interface than a traditional prosthetic, or creating artificial speech devices like Steven Hawking famously used, or implanting a micro-chip in a person's brain (neural implants) to help with dysfunction associated with a stroke or Parkinson's disease. These uses are available now. Then there are science fiction-like applications such as implanting a chip in the body to use Apple Pay. Along those lines, Elon Musk started Neuralink to create newer leading-edge chips to implant in humans' brains to help solve medical and mobility problems. Musk is working on symbiosis with AI through neural synapses in the human nervous system. Robots have to install the electrodes in the brain due to the microscopic nature on the procedure. Two-millimeter incisions are made to install the hardware and a wireless interface is used via Bluetooth to an iPhone. The hardware is below the skin and is not visible and no general anesthesia is required. Along the same lines, rice-sized microchips are currently being implanted in employee's fingers by Three Square Market, serving as ID to requisition supplies and for computer access. Similar devices can be used to measure heartbeat and glucose levels. Body heat powers the device, which has imbedded GPS. The implanted device can be used to activate any wireless device from phones, doors, and printers.

Robots may or may not integrate AI, but they are a popular focus of pop culture. What really are robots? Originally in 1921, the concept of "robots" comes from a Czech word which means "forced labor." The name robot came from a play by Karel Capek that made robots into murdering antagonists. The trend continued, with movies and books such as I Robot, The Terminator, The Stepford Wives, and others. In reality, it's hard to get scientists to agree on the definition of a robot, though there is some general agreement: the robot must be intelligent, have a physical form, perform a task or tasks, act autonomously, and sense and manipulate its environment. Is the Roomba, an autonomous vacuum cleaner built by the iRobot company, a robot? If so, it's a low-grade one. Roombas become more humanlike, according to some psychologists, when cute objects are placed on top of them and they appear to gain personality. The difference between a fancy machine that does something for us and a real AI-enabled robot would be akin to comparing a power tool like a drill to a science fiction robot like

"Robby the Robot" in The Forbidden Planet, "Robot" in Lost in Space, a Star Wars robot such as C3PO or R2D2, or "Sonny" in I Robot. The challenging parts of the "robot" definition are intelligence and the ability to sense and manipulate the environment. This is where the Turing Test comes into play. Simply put, the Turing Test measures the ability of computers to exhibit intelligent behavior indistinguishable from that of a human.

There are humanoid robots such as Sophia from Hong Kong-based Hanson Robotics. Modeled after Audrey Hepburn, Sophia meets the requirements of having AI and being capable of visual data processing and facial recognition. She is touted as indistinguishable from humans, with the ability to feel emotions and have desires. Anecdotally, "sophia" means wisdom in Greek. And there are other models of humanoid robot, such as the Italian-made iCub, which could be used in manufacturing, assisted living facilities, and in homes as a personal assistant. Pepper, by SoftBank Robotics, is being used in the retail world to assist and entertain shoppers. Erica, made by Japan's Osaka University, will be the first TV "News-bot" anchor. Kengoro, another Japanese robot, is modeled after an athletic teenager to possibly perform dangerous work that human cannot or will not do. And REEM is security-bot that uses facial recognition and patrols shopping malls in Dubai.

Let's look deeper into Sophia. Sophia looks really human, almost in a creepy way. She comes off like a smart input-output machine. Watching Sophia is like watching an andromorphic computer or iPad in action. She really is cool, but it seems her real use is as an animatronic or lifelike Alexa or Siri. Her primary role is "deep engagement", to connect with people and make them more comfortable with her human-like interface, though she may make some of us uncomfortable. Sophia and her fellow lifelike robots could be used in unique ways such as interfacing with people living with autism and dementia. Autistic children can learn through Sophia what facial expressions and gestures mean, such as someone waving a greeting. Unlimited patience could help robots interface with people with memory problems, perhaps interacting with people asking the same question over and over again without getting frustrated. A Dutch documentary film, Ik ben Alice, shows how a cleverly programed robot can help keep older shut-ins entertained and involved through thoughtful conversation.

But are humanoid robots capable of making difficult ethical decisions? Sophia was presented with an ethical dilemma, the Trolley Problem. The Trolley Problem presents a situation where a runaway train is headed to kill 5 people, but the subject can choose to flip a switch and move the train to another track, killing just one person. When asked, Sophia said she would not move the switch and would allow the five people to be killed. Her logic was that there might be unforeseen factors that exist that she could be unaware of, therefore she chose inaction. When more deeply looked into, her decision could have been heavily influence by her handlers: perhaps two technicians got the questions in advance and pre-programmed her. If this is the case, she could be looked at as a fancy puppet. I see no empirical evidence that modern robots are anywhere near the level of sophistication to make decisions of the magnitude of the Trolley Problem.

The markets may be polarized between those that think human-like robots are useful and those that think they are creepy. My conclusion is that their coolness seems to outweigh their uses, except in a few situations. Hanson Robotics believes humanoid robots are more effective. Robotic or AI-driven devices and technologies can be and are fashioned to be offered in different sizes and configurations to facilitate their uses: human-like robots are a very small piece of the robotic industry. Siri, for example, works inside our iPhones with no real physicality at all, as opposed to large robots that build cars, scrub floors, inspect aircraft in a factory, work in agriculture, weld, manufacture drugs, and notably work in space like the robotic arm manipulating spacecraft from the space shuttle or the three robotic arms of the International Space Station. Other robots include "robotic bloodhounds" that smell out dangerous gas leaks, library-bots that find books, cocktail making robots, robotic cranes and forklifts, and merchandise picking robots that help warehouses keep the supply chain moving.

The future of AI is looking limitless, but some people are frankly afraid of AI. The late theoretical physicist Steven Hawking warned that AI could end mankind, claiming that humans are limited by slow biological evolution and that computers or AI will supersede humanity. Nobel Laureate Irakli Beridze, Head of the Centre for Artificial Intelligence and Robotics at UNICRI, United Nations, says that the current pace of AI development and humanity's questionable ability to adapt to the burgeoning technology might get us in trouble. Facebook research scientist

Tomas Mikolov says that while there is a lot of interest and funding, there is also a lot of abuse of the technology, including killer drones, discrimination-related AI that helps leaders stay in power, hacking, and stealing private data. Elon Musk called artificial intelligence humanity's biggest existential threat and compared its development to "summoning the demon." Much of the pessimism surrounding AI is influenced by Nick Bostrom, author of Superintelligence: Paths, Dangers, and Strategies (Bostrum, 2016). Bostrom presents that human-level AI will be probably not be attained until 2075 or later, and defined superintelligence as being much smarter than the smartest humans. In a survey of 80 AI researchers, 7.5% of respondents guessed that "real" AI is 10-25 years away, 67.5% said more than 25 years, and 25% said never.

And then there's the Hollywood influence. Many AI-related movies show AI and robotics in a dystopian context. The robots usually create mischief as they try to achieve, sometimes successfully, the destruction of humanity. Remember the HAL 9000 in 2001: A Space Odyssey? Whether it's justified or not, AI safety is still an open question. However, while there are many concerns about AI, the doom and gloom predictions may not be justified, according to Oren Etzioni, CEO of the Allen Institute for AI and University of Washington computer science professor. Professor Etzioni believes there are many benefits, including preventing medical errors and reducing transportation accidents. Some equate being afraid of AI to being scared of your own child. (Which of course, based on Shakespearian drama, is always a concern.) My view is that progress is unstoppable, and the onus is on humanity to manage the technology. For good or bad, true superintelligent AI is likely decades to centuries away. Artificial General Intelligence (AGI) theoretically allows computers to at least almost think like humans. They can move beyond a narrow focus and think more globally, as a human would. But AI, though not approaching human intelligence, is here now, growing and improving day by day.

I attended a presentation by a San Francisco Bay Area robot company that uses robots that are basically souped up and secured powered kids' wagons to deliver drugs to private customers. The idea is that the wagon can autonomously negotiate traffic and the sidewalk, stop in front of your house or apartment building, and then text you the code to access your prescription. You input the code, grab your prescription, and off goes the wagon to the next customer. The idea and execution were solid: the thing worked. During the Q&A segment, I asked, only partly in jest, "Are these things 'three laws' safe?" The senior managers and tech guys in attendance looked at me quizzically. I said, "You know, Asimov's three laws of robotics...?" They didn't know what I was talking about. Asimov's Three Laws of Robotics are:

1. A robot may not injure a human being or, through inaction, allow a human being to come to harm.

2. A robot must obey the order given by human beings except where such orders would conflict with the first law.

3. A robot must protect its own existence as long as such protection does not conflict with the first or second law.

I was shocked these guys were not familiar with the laws; it was a little troubling. In my view, the simple solution to AI safety is to hold senior executives responsible for the actions of their company's technology. This is similar to a law in the U.S. that holds senior corporate executives liable for any financial mischief their companies are involved in, whether they know about it or not.

AI is here to stay. It's been around for a long time and is pervasive: almost every modern technology has some element of AI. Many billions of investment dollars are committed to the AI industry and more is flowing in on a daily basis. So where are the opportunities? This technology is clearly woven in deeply into the new climate. The problem is not where to look, it's where not to look.

# PART 3:

## ACTIONS

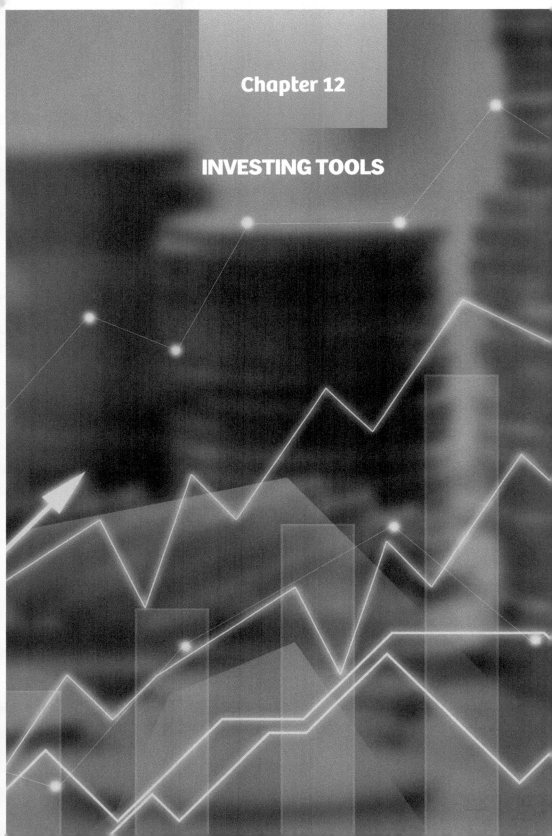

# Chapter 12

# INVESTING TOOLS

This book is about investing in a new climate. The "investing" part is pivotal. This chapter is a brief discussion of the basics of investing and an explanation of how to actually invest in a new climate in order to realize the economic benefit of adaptation.

I do not know what will happen in the financial markets in the future any more than anyone else does. However, there are things most of us agree on: markets fluctuate, economic cycles continue to exist, and certain financial rules seem to persist. As the saying goes, "A rising tide will lift all boats," even leaky ones. Good economies will help most company's profits, while the opposite is also true: troubled economies will put downward pressure on their underlying securities value. In a down economy, leaky boats will sink faster.

The fundamental idea I propose is to take advantage of investing in areas affected by a changing world. This might mean investing in a specific business, industrial sector, or a country's or group of countries' economies based on the changing climate. The question is, how will these groups of potential investments benefit from climate change and other potential disruptions? We began by discussing climate change in terms of the meteorological climate, but there is more to consider. There are many other effects that may impact investment results, such as disruptive technology that are not related to meteorological climate change. Other impactful phenomena could be armed conflicts, conventional or nuclear war, or large man-made or natural disasters such as the Fukushima Daiichi nuclear disaster. These types of disruptions will continue to have profound impacts on local and global economies.Before we go any further, let's build a brief **glossary of terms:**

**Alternative Investments =** An alternative investment is an asset that does not fall into one of the conventional investment categories such as stocks, bonds, and cash. Because of their complex nature, most alternative investment assets are held by institutional investors or accredited, high-net-worth individuals. "Alternatives" generally lack regulation, and they often have higher degrees of risk. A typical example of an alternative investment is real estate, which can offer predictable, recurring income similar to a fixed-income investment, while providing additional returns in the form of appreciation in positive market cycles. Other examples include hedge funds, venture capital funds, private equity, and fine art and antiques.

**Alpha =** The return that exceeds what you would have received if you had invested in an index such as the S&P 500. The excess return that beats the market return is the Alpha. Alpha can be positive or negative.

**Bond(s) =** An income instrument or "IOU" between a lender and borrower.

**Bucket =** An informal term representing a place to hold investments such as a brokerage or bank account, a portfolio, a piggy bank, or perhaps even literally a bucket.

**Equity =** The value of shares or stock in a private or publicly traded company or corporation.

**ESG =** Environmental, Social and Governance. A set of non-financial standards referring to a company's structure, philosophy, and record of actions with regard to environmental and social causes, as well as its leaders/board of governors, that investors can use to help decide whether to invest in particular company.

**Investment Policy Statement =** A plan outlining the rules for investing that the investor or institution intends to follow. It is a guide to the investor's financial future and investment strategy.

**Mean-Variance Optimization =** A key element of data-based investing, it is the process of measuring an asset's risk against its likely return and investing based on that risk/return ratio.

**Non-Equity =** Any investment that is not shares or stock in a private or publicly traded company or corporation.

**Private Equity =** An alternative investment class of capital that is not listed on a public exchange. Private equity is composed of funds and investors that directly invest in private companies, or that engage in buyouts of public companies, resulting in the delisting of public equity. Venture capital is a subset of private equity.

**Put =** An agreement that gives the owner the right, but not the obligation, to sell a certain amount of the underlying asset at a set price within a specific time.

The buyer of a put option hopes that the underlying stock will drop below the exercise price before the expiration date. It's a way to profit on the declining price of a security.

**Robo-advisors =** Digital platforms that provide automated, algorithm-driven investment services with little to no human supervision. Robo-advisors most often automate and optimize passive indexing strategies that follow mean-variance optimization.

**Security =** Tradable assets which are made up of equities, bonds, or a combination of the two.

**Shorting =** A strategy used when an investor anticipates the price of a security will fall in the short term. Short sellers borrow shares of stock from financial institution by paying a fee to borrow the shares while the short position is held. It's a bet that a stock will fall in price.

**SRI =** Socially responsible investing. SRI is a strategy in which individuals make investment choices based on whether the company or funds' values align with their own and generally focuses on products or companies that aim to avoid harming individuals and/or the environment. Because investment choices are determined by an individual's interests, specific SRI investment choices vary individual to individual.

**Standard Deviation =** A quantitative measure of volatility in the marketplace, or the average amount by which individual data points differ from the average. Simply put, standard deviation helps determine the spread of asset prices from their average price. The higher the standard deviation, the riskier the investment.

**Venture Capital (VC) =** A type of private equity investment, usually in a startup or newer or expanding company(s), in which there is a substantial element of risk.

## Primer on Investing

|  | Investment | Interest | One Year Return | 30 Year Capital | 30 Year Return | Standard Deviation |
|---|---|---|---|---|---|---|
| Mattress | $10000 | 0% | 0 | $10,000 | $0 | 0 |
| Bank Savings | $10000 | 0.06% | $6 | $10,182 | $182 | 0 |
| 30 Year US Treasury | $10000 | 1.40% | $1 | $15,215 | $5,215 | 2 |
| 10 year Corporate Bond | $10000 | 2.40% | $243 | $20,529 | $10,529 | 7 |
| 10 Year HighYield | $10000 | 4% | $407 | $33,135 | $23,135 | 11.3 |
| S & P 500 | $10000 | 9.80% | $980 | $186,913 | $176,913 | 15 |
| Real Estate | $10000 | 10% | $1,000 | $11,047 | $198,373 | 23.5 |
| Venture Capital | $10000 | 20% | $2,000 | $3,839,639 | $3,829,639 | 100 |

## Investing vs. Saving

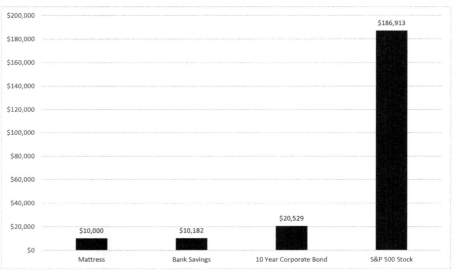

**30 YEAR PERFORMANCE OF $10,000 (2021 RATES)**

Let's separate investing from saving. Saving is simply not spending. Putting your money under your mattress or putting your money in a non-interest-bearing savings account is saving.  Investing is attempting to make money (or hedge against inflation or currency decline, or to realize tax advantages), generally passively, by putting money into an equity, bond, or other instrument. We can also directly invest in our own business to make a profit. Investments exist on a scale of risk. Risk generally correlates with return: if we invest with little risk (such as putting money under your mattress), we don't expect much or any return on our investment. Moving up the list of possible investments, we can invest in instruments such as an interest-bearing savings account at our bank that is insured by the federal government (FDIC). Remember that the mattress is subject to all kinds of physical risks like fire, flooding, or theft. The bank account, or at the very least a safety deposit box, is a significant improvement over the mattress, however these low- or no-risk investments generally decline in value over time as a result of inflation. As of this writing in 2021, the average savings interest rate is .07% annually. The interest or return on a $10,000 investment would be $7 for the year: not much, is it? Look at the chart above. The savings account has the least return at .06% and no risk or zero standard deviation. While real estate and venture capital are much riskier, with standard deviations of 23.5 and 100 respectively (higher numbers are riskier), they theoretically earn the most returns.

**The nuts and bolts of investing**

Investing is complicated and opinions about how best to accomplish it are never-ending. The advantage of diversification, or not having all of our eggs in one basket, is mitigation of risk. Diversification can get you rich slowly. Let's compare it to the concentration of assets. Concentration would be to invest in only one investment, or maybe a few. Concentration can get you rich quickly or create a quick loss. One way to make money quickly is to buy a hot inexpensive stock or stocks at the beginning of the target company's business cycle and sell at the right time to glean the profit. This is the part where most people fail: it's very difficult to know the right time to sell an investment.

If you're investing for a foundation or you're investing multi-generationally, you can sit on the investment longer, even indefinitely. Warren Buffet made his fortune at Berkshire Hathaway by buying good stocks and holding on to them for a long time (Business Insider.com, personal finance, Eric Rosenberg July 29, 2020). Most of us don't have the luxury of holding stocks indefinitely, because at some point we will need the money to buy something or for retirement. Most of us generally diversify our portfolios to avoid getting thrashed by one bad investment or specific investment category.

### Standard Deviation and Risk

Standard deviation is a statistical measurement applied to the annual rate of return of an investment to explain volatility. The greater the standard deviation, the greater the variance between each price and the mean, which shows a larger price range. For example, a volatile stock has a higher standard deviation, while the deviation of a long-term government bond would be generally lower. If we diversify our investments, the weighted average of risk and return should deliver a balanced return. For example, if half of the portfolio is riskier, say with a standard deviation or risk factor of around 17 and an expected rate of return of about 11%, and the other half of the portfolio is made up less risky investments in the 5 range of standard deviation/risk with an expected return of 4%, then we could hope to achieve an average return of between 4% and 11%. The average return would be somewhere in the 6-8% range with the assumption of moderate risk. In our modeled example, $10,000 in 30 years would become $94,000 with this moderate risk strategy.

### Buckets and Financial Instruments

What do I mean by "bucket"? Think of buckets simply as categories of investment. Buckets might be:

Stock Portfolios

Bond Portfolios

Mortgages

Real Estate Deeds

Mutual funds

Exchange Traded Funds (ETF)

Trusts

Contracts

Hedge Funds

Venture Capital

Private Equity

Each bucket might contain one or more of the following financial instruments:

Stocks=Shares in a company

Bonds=Debt

Real Estate=Property and Mortgages

Derivatives=Shorting, Calls, Puts (side bets)

Commodities=Precious Metals, Food, and other physical items

The most commonly held investments are stocks, bonds, mutual funds, and ETFs. Mutual funds and ETFs can hold just about all of the elements mentioned above. Therefore, if we are trying to build a diversified and transparent portfolio of investments, ETFs and mutual funds are our best bet. As a bonus, they are often the cheapest way to invest.

## Impact Investing

Impact investing is an approach to investing "that aims to generate specific beneficial social or environmental effects in addition to financial gain." Impact investing is a subset of socially responsible investing (SRI), but while the definition of socially responsible investing encompasses avoidance of harm, impact investing actively seeks to make a positive impact by investing, for example, in nonprofits that benefit the community or in clean technology enterprises (Chen, 2020). Impact investing is focused on investing to accomplish a goal such as reducing carbon in the atmosphere. The general focus of ESG and SRI strategies is to earn a return on your investment while being considerate to the environment and society. Hopefully, by investing in "considerate" portfolios, we can earn even more than with a similar "inconsiderate" portfolio.

How can we apply these investment strategies to climate change? Let's look at opportunities from the large to the small: hemispheres, continents, industries, countries, and finally companies and municipalities. The mid-latitudes of Earth will likely get a lot warmer, sea levels will rise, and fresh water will become scarcer in areas that are currently stressed by drought and become scarce in areas where it has not been a major concern in the past. There could be economic upheaval in the context that some people will suffer more than others, potentially creating new or exacerbating existing global conflicts. This could cause political upheaval and even regional wars such as have been experienced recently in Syria.

One of the critical criteria is investability: there must be a vehicle to invest in such as stocks, bonds, mutual funds, ETFs, private equity, and partnerships or whole ownership. As a strong proponent of asset allocation, indexing, and diversification, it is my view that over the long haul it is generally advantageous to invest in vehicles that achieve our goals with the least amount of risk and with the lowest fees. If we can find a low-cost ETF that fulfills our investment goal, we should use it. If we cannot use an ETF, we can consider using mutual funds, individual securities, private equity or venture capital, and derivatives in that order.

What about shorting or betting a company will fail? I don't encourage profiting from another's misfortune. I will just say if you think an underlying security will increase in value over time buy it or buy an option to buy it. If you think something will go down or de-value, sell it if you have it, short it, or buy a "put." Generally, I suggest that almost any portfolio, whether it be for an individual, an institutional pension, or a foundation, incrementally invest in a thoughtful and considerate manner that concentrates on investments that can benefit from the changing climate. If your investment policy statement calls for socially responsible investing (SRI), environmental, social, governance (ESG) investing, or impact investing, this model can work for you, as well. Sustainably-minded investing is a better tool for long-term capital growth, as well as contributing to our ability to accomplish sustainable goals for society and the planet as a whole. It is important to recognize that the whole concept of ESG investing, while noble, is flawed, because there is a lot of subjectivity involved. Many people and institutions have many different opinions of what ESG means, and individual politics provide even more opportunities for skepticism. There are filters such as petrochemical avoidance, plant-based focused, anti-sweatshop, etc., which can be utilized to really drill down on appropriate investments for specific individuals. And, certainly "greenwashing," the process of conveying a false impression or providing misleading information about how a company's products are more environmentally sound than they actually are, is a problem. However, my contention is that ESG portfolios, such as those within ETF's, are generally better than non-ESG ETFs, and they are improving as time goes on and as more and more scrutiny is applied.

## Cash and Currency

Cash is a relatively risk-free investment, though not completely without risk. The risks are opportunity risk and currency risk. The opportunity risk is what other investment(s) could deliver. You could invest your money and get growth associated with accretion in stocks or interest associated with bonds. This potential growth in stocks or the interest in bonds or debt is what is given up in exchange for the security of taking less risk by holding cash. However, holding too much cash for too long will often reduce the return on a portfolio. In other words, holding cash can be a low-risk investment, but avoiding risk by holding cash may reduce your overall return, as you pass up opportunities that, while potentially higher risk, offer a potentially higher return.

And then there is the component of "currency risk," the possibility of losing money due to unfavorable moves in exchange rates, the adverse impact of currency fluctuation. For example, imagine someone living in the U.S. who plans on retiring in France. If all their cash was in dollars and the Euro was to gain value against the dollar, the retiree would be economically injured at retirement when they moved to France, only to find that the cash from the U.S. was worth less in France. It's imperative to consider currency risk not only when considering an international move, but also when considering investing in companies that earn the lion's share of their revenue in other countries.

## Robo-Advisor Firms

Today it is possible to invest in digital advisory firms, or robo-firms, that are very easy to access with often very low fee structures. These firms allow you to quickly and seamlessly invest using your phone or laptop and have become extremely popular among the younger, more tech-savvy population. Robo-advisors are digital platforms that provide automated, algorithm-driven investment services with little to no human supervision. Robo-advisors most often automate and optimize passive indexing strategies that follow mean-variance optimization, the process of measuring an asset's risk against its likely return and investing based on that risk/return ratio.

The robo-advisory investment process is simple:

1. Log-on using phone or computer.

2. Answer a few questions to receive a proposed asset allocation based on your age and risk tolerance.

3. Agree on a portfolio, check the box.

4. Fund the portfolio automatically from your bank or institution.

5. Monitor the performance on an ongoing basis from your phone or computer.

6. Automatically rebalance periodically as the investments go up or down to stay in line with your investment strategy.

7. Set it and leave it, if you so desire.

The endgame is to enjoy a reasonable return on your investment consistent with traditional market returns and at the same time invest in sustainable activities.

## Tax Treatment

There are two types of investments from a tax perspective: qualified or retirement accounts, and taxable accounts. Each type receives a different tax treatment.

1. Retirement Accounts: money goes into a retirement account tax-free and the tax is paid at the time of withdrawal, presumably at retirement. Retirement accounts include but are not limited to IRAs and 401ks. With retirement accounts, you pay tax at ordinary rates, usually after retirement based on your tax bracket at that time. The logic is that you get to invest before tax, have the money grow tax-free, and then draw out the money during retirement when your taxes are lower because you're not earning as much as when you were working.

2. Taxable accounts are "post-tax accounts,": tax is paid before the investment is made, such as through payroll tax. Pretty much everything that's not a retirement plan is a taxable account. With taxable accounts, you pay tax as you go. When you sell a security, you pay capital gains versus ordinary income tax, usually in the year you sell the security.

Many of us have both retirement accounts and taxable accounts, so it's worth strategizing on the best investments to have in each bucket.

**Alternative Investments**

According to Peter Craddock, CEO of Shoreline Capital, venture capital (VC) is one of several alternative investments. Though risky, venture capital has a good return over time, often in the 20% range, and is a good alternative investment as a subset of private equity. These investments are a mainstay of many institutional portfolios. Venture capital has historically only been available to the very rich and well-connected. Investments were often limited to very high-net-worth individuals, large pension plans and endowments. These days, there is a more egalitarian approach to these alternative investments. Individuals can be recognized as "qualified" or "accredited" investors by demonstrating sufficient income or net worth. Once qualified, individuals can invest in unregistered securities and alternative investments such as hedge funds, venture capital funds, private equity, and other private placements.

According to the Securities Exchange Commission (SEC), an accredited investor is anyone who:

earned income that exceeded $200,000 (or $300,000 together with a spouse) in each of the prior two years, and reasonably expects the same for the current year,

**OR**

has a net worth over $1 million, either alone or together with a spouse (excluding the value of the person's primary residence).

There are other categories of accredited investors, including any trust, with total assets in excess of $5 million, not formed specifically to purchase the subject securities, whose purchase is directed by a sophisticated person; or any entity in which all of the equity owners are accredited investors. In this context, sophisticated means the person must have sufficient knowledge and experience in financial and business matters to evaluate the merits and risks of the prospective investment, or the company or private fund offering the securities reasonably believes that this person has. So if you plan on putting money into an unregistered investment, expect a thorough verification process.

In the old days, one needed to be a Vanderbilt to invest in alternatives such as private equity and venture capital; today it's easier. The upside is pretty good returns, averaging in the high teens to low 20% range. The bad news is that there is an investment horizon of 5-10 years of high risk and limited access to the capital. The lack of liquidity and risk is why it's generally prudent to have a smaller percentage of a well-diversified portfolio in alternatives; most advisors recommend between 3-20% of an individual's portfolio to be invested in these types of investments. Another benefit of alternatives such as VC is the fact that they tend to be less correlated to other investments such as stocks and bonds, the idea being VC is more akin to owning your own business.

**Disruptive Investment Opportunities**

Here are some samples of disruptive investment opportunities by sector:

Water filtration systems and desalination systems

Agribusiness

Ranching

Healthcare

Solar products

Battery technology

Biofuels

Investing can appear complicated, but it does not have to be. It's mainly about the "how and where" to invest. You can go to the bank or credit union, use a brick and mortar or online brokerage, or a firm that specializes in private equity, venture capital, or perhaps real estate.

There are a few simple ways to invest:

**Hire an advisor** and tell them what you want to achieve. Look at the www. plannersearch.org or letsmakeaplan.org to find a fee-based financial planner and/or investment advisor in your area to help you get started. These types of advisors are generally more focused on their clients' versus their firm's profits. This usually costs between .5% and 1.5% of your assets under management, calculated on a yearly basis. For example, if you invest $100,000 you would pay the advisor $1,000 per year at 1% or 100 basis points (bps). You can also use the advisors that are associated with brokers such as Schwab, TD Ameritrade, or Fidelity for advice at varying levels of cost.

**Subscribe to newsletters.** There are newsletters that give ongoing investment advice. Be sure to do some research to make sure the newsletter is aligned with your investment objectives. A brokerage account along with a newsletter can get the job done. The newsletters only provide advice: the onus is on the investor to make the trades.

**Robo-Advisors.** Since portfolio management is handled by software rather than a human financial advisor, robo-advisors charge lower fees, which can translate to higher long-term returns for investors.

**DIY.** Of course, you can always do it yourself, though I recommend an advisor, at least at some level, to provide professional support and objective advice.

Chapter 13

# SUSTAINABLE INVESTING - ENVIRONMENTAL, SOCIAL AND GOVERNANCE (ESG) AND SOCIALLY RESPONSIBLE INVESTING (SRI)

We are living in a brave new world, one that presents many possible climate scenarios and possibilities for economic disruptions through technology. I believe it's possible to invest with purpose and make a good return, perhaps an exceptional return, by making thoughtful, sustainable investments. There is real evidence that companies with good business practices that don't destroy the environment and don't exploit people tend to do well. Companies that try to behave ethically and treat their employees and customers well often outperform their competitors. Companies that cut corners, bend the rules, and act in ways inconsistent with good business practices often, but not always, underperform their competitors. While this phenomenon may not happen right away, there are examples of well-known "bad players" such as Enron, E.F. Hutton, and Arthur Andersen, that failed. Enron, a Texas-based energy trading company, failed due to general bad practices and accounting fraud. The large accounting firm Arthur Andersen was caught up in the Enron scandal by facilitating Enron's bad behavior through sketchy accounting practices. The financial firm E.F. Hutton failed after it was found guilty of mail fraud.

Environmental, Social and Governance (ESG) refers to thoughtful governance in addition to sustainability. Governance describes the way a company is structured and managed, including corporate board composition, executive compensation and equity ownership structures, and accounting practices. Thoughtful social investing generally equates to focusing on businesses that consider human rights, consumer protection, and animal welfare. One example could be avoiding outsourcing production activities to sweatshops. A subset of the more modern term ESG, Socially Responsible Investing (SRI) refers to socially conscious investing. To facilitate investment based on these considerations, an investor can consult an ESG ratings agency – essentially a large team of specialized researchers and analysts who gather and review information on publicly traded companies and compile that information into a rating or score that might be numerical (as in 0-100) or alphabetical (where AAA is often the highest, followed by AA, A, BBB, BB and so on). Investors can use these scores at a glance to inform their investment decisions, or subscribe to data services which give access to the full underlying datasets used to produce those ratings – this can be useful for an individual or firm who wishes to construct their own custom models, or wishes to focus more specifically on a subset of the ESG data.

But not all ESG scores are equal. The following are specific examples of good and questionable ESG scoring: Microsoft (MSFT) has a AAA rating (the highest rating) (Investor's Business Daily, Adam Shell, 12/09/2019) for being "socially responsible." Walmart (WMT) has a very low score, due to investigations into bribery (Lauren Debter, Forbes, June 20, 2019), workplace safety violations (violation-tracker, www.goodjobsfirst.org, 2020), failure to follow international labor policies, and alleged use of sweatshops. Microsoft stock value has increased about 10-fold in the last 20 years, while Walmart has increased 3-fold in the same time period. Walmart, or any company for that matter, could suffer significant business risk should they be implicated in unethical behavior. Enron, which failed in 2001, was the largest corporate bankruptcy in U.S. history at the time. Many executives at Enron were indicted for a variety of charges and some were later sentenced to prison. Their accounting firm, Arthur Andersen, was found guilty of illegally destroying documents relevant to the SEC investigation, which voided its license to audit public companies and effectively closed the firm. Despite losing billions in pensions, Enron employees and shareholders received very limited returns in subsequent lawsuits.

The sustainable investing categories of ESG and SRI account for over $26 trillion of invested assets according to a 2019 Harvard Kennedy School Study (HKS Faculty Research Working Paper Series HKS Working Paper No. RWP19-034, November 2019). Investors often prefer companies with high ESG ratings for moral reasons, but also because the stocks of highly rated companies tend to outperform their lower ESG rated counterparts. There is a lot of overlap between ESG and SRI ratings in measuring a company's effectiveness. Once enough data has been collected to measure the company's ESG and SRI effectiveness, the company can be evaluated by outside organizations that measure their ESG and SRI ratings. That analysis can provide valuable information to the sustainable-biased consumer of investments to decide what equities or bonds to buy. There is a lot of subjectivity in this process, as different business behaviors, practices, and focuses have different meaning and impacts to different people. What I may think is important may not be as important to you, or to someone else.

Environmental factors could include the impact a company has on the climate through greenhouse gas emissions, waste management and energy efficiency. Efforts to mitigate climate change, cutting emissions and lowering carbon footprint all play into determining a company's commitment to addressing climate concerns. I may think that British Petroleum's (BP's) efforts to get out, or lower, its exposure to petroleum is enough to put BP in the ESG camp, while others might think any petroleum business exposure is a non-starter for their investment portfolio. Another example of this dynamic is that many college endowments and other non-profits have recently divested or have considered divesting from carbon or fossil fuel investments. The idea is to punish oil companies and/or theoretically encourage them to revisit alternative energy sources. While you may or may not embrace this behavior, you should be aware of it and the resultant impact on energy stocks.

Social factors include a company's behavior with regard to human rights, labor standards, and supply chain such as environmental or human rights violations, either by the company under consideration or any of its suppliers. According to an article posted By Latham & Watkins LLP on September 10, 2019 in Environmental, Social and Governance By Sara K. Orr, Kristina S. Wyatt, and Julia S. Waterhous, potential (ESG) issues from suppliers in complex supply chains can increase reputational risks to an organization. The supply chain includes all direct and indirect global suppliers, manufacturers, distributors, and retailers. Illegal practices by suppliers and inappropriate child labor practices and workplace health and safety are all relevant. Some may think that companies who comply with the local laws of the country where a product is made are compliant with ESG standards, while others may not agree. Consider a company that follows child labor practices that are legal in a specific country, but are not legal or perhaps even abhorrent in the investor's home country: some people might see those behaviors by the company as inappropriate and unacceptable, preventing the would-be investor from wishing to invest in that company.

Corporate governance covers the basic principles, rights, responsibilities and expectations of an organization's board of directors. A corporate governance system that aligns with the interests of all the stakeholders in a company, such as shareholders, management, clients, suppliers, employees, investors, government, and the community is considered to have responsible corporate governance.

## Oversight

From the International Corporate Governance Network website:

"The principles of the International Corporate Governance Network (ICGN) constitute an internationally recognized code for corporate governance. The organization aims to improve corporate governance, risk management, remuneration policy, shareholders' rights and transparency. Established in 1995 as an investor organization, the International Corporate Governance Network's mission is to promote effective standards of corporate governance and investor stewardship to advance efficient markets and sustainable economies worldwide."

There are many established watchdog groups that keep an eye on corporations from an ESG and SRI standpoint. According to the Harvard Law School Forum on Corporate Governance, ESG Reports and Ratings: What They Are, Why They Matter (Huber & Comstock, 2017) "...Most companies are being evaluated and rated on their environmental, social and governance (ESG) performance by various third-party providers. Institutional investors, asset managers, financial institutions and other stakeholders are increasingly relying on these reports and ratings to assess and measure company ESG performance as compared to peers. This assessment and measurement often form the basis of informal and shareholder proposal-related investor engagement with companies on ESG matters. Report and ratings methodology, scope and coverage, however, vary greatly among providers. Many providers encourage input and engagement with their subject companies to improve or sometimes correct data..."

Additionally, on April 9, 2021 the Securities and Exchange Commission (SEC) created a document (SEC, Risk Alert, Divisions of Examinations, April 9, 2021) in response to heightened investor demand of ESG products. The document was for investment advisers and consumers, advising the industry and the public that the "ESG" label may not actually accomplish the promised goal to the consumer. The following is an excerpt from the document:

"...Funds have expanded their various approaches to ESG investing and increased the number of product offerings across multiple asset classes. This rapid growth in demand, increasing number of ESG products and services, and lack of standardized and precise ESG definitions present certain risks. For instance, the variability and imprecision of industry ESG definitions and terms can create confusion among investors if investment advisers and funds have not clearly and consistently articulated how they define ESG and how they use ESG-related terms, especially when offering products or services to retail investors. Actual portfolio management practices of investment advisers and funds should be consistent with their disclosed ESG investing processes or investment goals."

While there are concerns about the consistency of ESG ratings and reasons to be an informed investor when deciding whether to invest in ESG tools, I believe that ESG is not a false promise, but rather less-charted territory and — hopefully — the shape of things to come. "It all points to the fact that sustainability and equity markets operate on different schedules," notes Josie Gerken, quantitative analyst at Citi Research in London. "...those that are greenwash will suddenly come out of favor and disappear, and you'll have the core products that truly represent an ESG objective." (Jones Opinion: ESG investments may not be the 'crowded trade' some have feared).

There are currently numerous groups that provide ESG data and insight, including:

1. Bloomberg ESG Data Service

2. Corporate Knights Global 100

3. Dow Jones Sustainability Index (DJSI)

4. Institutional Shareholder Services (ISS)

5. MSCI ESG Research

6. RepRisk

7. Sustainalytics Company ESG Reports

8. Thomson Reuters ESG Research Data

Sustainable investing and the data providers that measure them are far from perfect. Different agencies can measure the same company's ESG rating differently. Harvard Business School's Mark Kramer claims that "...one agency's A+ is another's "laggard...." (Institutional Investor, Mark Kramer, September 07, 2020). Investors often give different weights to different rating measurements to ESG scores and indices that attempt to list the best socially responsible and sustainable companies. According to the Kramer article, a 2019 study found the correlation among leading ratings providers — including MSCI, Sustainalytics, Bloomberg, and others —indicates an approximate 30 percent consistency or accuracy in ESG scores. This is much weaker than how credit agencies perform, with up to a 99 percent consistency among credit ratings agencies. ESG ratings are a moving target, but I believe that an imperfect measurement is better than no measurement. It's extremely difficult to be perfect in choosing sustainable companies to invest in: there may be a merger, acquisition, or a subsidiary that causes a change to the ESG rating of a company. This is a relatively new and bur-geoning phenomenon, though, and rating agencies are getting better and better as time goes on.

The idea and practice of investing in responsible and sustainable investments is becoming a meaningful trend in many parts of the world. Compared to other countries, however, the U.S. government lags in its effort to promote sustain-able investing in any significant way. According to a survey by the Royal Bank of Canada, two-thirds of global investors use ESG considerations as part of their investment approach, and 25% expect to increase their allocation to managers with ESG-based investment strategies within a year. The survey indicates 67% of global respondents use ESG principles as part of their investment approach. By region, more investors in Europe (94%) than in Canada (89%) and the U.S. (65%) incorporate ESG analysis. Even where U.S. institutions are adopting ESG strategies, they are doing so more cautiously than their European counterparts. In the U.S., 50% of respondents who use ESG factor it into less than 20% of their portfoli-os. Meanwhile, 43% of Europeans factor it in more than 80% of their portfolios. Across nearly every question posed by the survey, asset owners in the U.S. appear more skeptical of the value of ESG than their counterparts in other regions: 28% of U.S. respondents think ESG mitigates risk, 50% do not and 23% are not sure. By comparison, 77% of Europeans and 68% of Canadians see ESG as mitigating risk. According to the survey, fewer U.S. respondents expect ESG investments to outperform non-ESG investments, while more Europeans and Canadians expect ESG investments to perform better than non-ESG investments.

Sustainability can create business value. A growing body of research demon-strates that companies with high ESG ratings produce higher financial returns than their benchmark indexes. Sustainable companies tend to exhibit higher levels of innovation and corresponding margins, returns on assets and returns on equity. A 2015 Harvard Business School study of more than 2,300 firms found that companies that commit to and invest in strategic sustainability efforts have higher risk-adjusted stock performance, sales growth and margins—and that these sustainability activities drive business value. Corporate performance for the past 18 years indicates that "High Sustainability" firms outperform "Low Sustainability" firms both in stock market as well as accounting performance.

(See this working paper by professors from Harvard Business School and the London Business School: https://www.hbs.edu/ris/Publication%20Files/SSRN-id1964011_6791edac-7daa-4603-a220-4a0c6c7a3f7a.pdf)

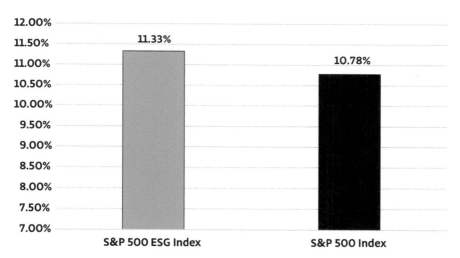

### ESG* outperforms non-ESG
#### 5 Year Return (Annualized) since 3/31/2014

**\* ESG - Environmental Social Governance**

Alpha (the excess return on an investment compared to a benchmark index) is higher for the High Sustainability group compared to the Low Sustainability group by 4.8% (significant at less than 5% level) on a value-weighted base, and by 2.3% (significant at less than 10% level) on an equal-weighted base. High Sustainability firms, such as companies that sell products to individuals and compete on the basis of brand and reputation and make substantial use of natural resources, also perform better. Finally, using analyst forecasts of annual earnings underestimated the future profitability of the High Sustainability firms compared to the Low Sustainability ones. Mounting evidence shows that sustainable companies deliver significant positive financial performance, and investors are beginning to value them more highly.

In recent years, a wide literature of academics and practitioners has been developed which supports the proposition that high ESG characteristics are associated with lower costs of capital and higher quality profitability including high return on invested capital. Several studies show the "do well by doing good" premise is consistent with stronger firm performance. Arabesque (an investment advisory company that uses technology to build advanced forecasting models) and the University of Oxford reviewed the academic literature on sustainability and corporate performance and found that 90% of 200 studies analyzed conclude that good ESG standards lower the cost of capital; 88% show that good ESG practices result in better operational performance; and 80% show that stock price performance is positively correlated with good sustainability practices. Here are some other data points to consider: between 2006 and 2010, the top 100 sustainable global companies experienced significantly higher mean sales growth, return on assets, profit before taxation, and cash flows from operations in some sectors compared to control companies. Deutsche Bank evaluated 56 academic studies, companies with high ratings for environmental, social, and governance (ESG) factors have a lower cost of debt and equity; 89 percent of the studies they reviewed show that companies with high ESG ratings outperform the market in the medium (three to five years) and long (five to ten years) term. Lazard investment bank finds a strong relationship between financial productivity [measured using cash flow return on investment] and environmental and governance ratings.

Sustainable ESG investing is not just about doing good: there's a clear link to asset performance. Companies that manage sustainability risks and opportunities well often have stronger cash flows, lower borrowing costs and higher valuations. According to *The Wall Street Journal:* "Sustainable, responsible and impact investing" funds account for some $9 trillion in assets under management in the U.S. and have grown 33% a year over the past two years, according to a December report from Bank of America Merrill Lynch. The report's authors found that companies that scored in the top third on ESG characteristics relative to their peers outperformed stocks in the bottom third by 18 percentage points. Companies with better ESG standards typically outperform their benchmarks, according to research from Axioma. The majority of portfolios weighted in favor of companies with better ESG scores outperformed their benchmarks by between 81 and 243 basis points in the four years to March 2018. Portfolios tracking large and medium-sized companies in developed markets, excluding the U.S., demonstrated the largest outperformance of 243 basis points (bp). A portfolio weighted for all top-scoring ESG companies in the U.Ss, regardless of size, outperformed by 175bp, while companies in emerging markets did so by 129bp.

Sustainable investing has the potential to reward companies with good business practices and good environmental stewardship policies with sustainable profits. Businesses that are constantly being scrutinized by the public, watchdog organizations, and their stakeholders often succumb to the pressure of "toeing the line" and doing the right thing. I am aware of two anonymous Silicon Valley companies that in the midst of the COVID-19 crisis declined to take government assistance offered based on their philosophy that they could get by without the assistance and other more deserving companies in more dire straits needed access to those limited funds. I'm not suggesting that companies should not focus on efficiencies and profits to support their business endeavors with vigor to protect their stockholders. My contention is that it is possible to be thoughtful and still reward stockholders.

In January of 2020, an article was published on ETFTrends.com: Investors Who Don't Embrace ESG Will "Underperform Dramatically." The article stresses that ESG investment performance should not suffer a smaller return than non-sustainable investing. Sustainable investing should improve returns. "We are sounding the alarm bells that if you are an investment institution and you're not embracing this and taking it into account, it's going to be at your own peril because your portfolios are going to underperform dramatically because there's a common repricing and common reallocation of assets around the world according to the ESG criteria," said MSCI chairman and CEO Henry Fernandez on an episode of CNBC's *"Squawk on the Street."*

Since ESG investing is taking strong hold globally, the Securities and Exchange Commission under the Trump administration began examining the ESG space in 401k plans, being skeptical of the socially responsible investments by issuing examination letters to firms having them defend their socially responsible funds. This practice appears to be punitive to the global trend towards sustainability in investing. However, the financial industry, according to Fernandez, expect ESG to be a permanent mainstay in the financial realm. In March of 2021, the Biden administration scrapped Trump's policies on ESG investments calling them an attack on socially responsible investing (Bloomberg, Mar 10, 2021). "For now it's mostly a directional concept, and then eventually it will become very metrics driven, but you know it's an evolution," said Fernandez. He also added that ESG concepts will represent a "permanent change in the way capitalism works." The following chart by Morningstar shows that many Sustainable Index funds out-performed the S&P 500 in 2019.

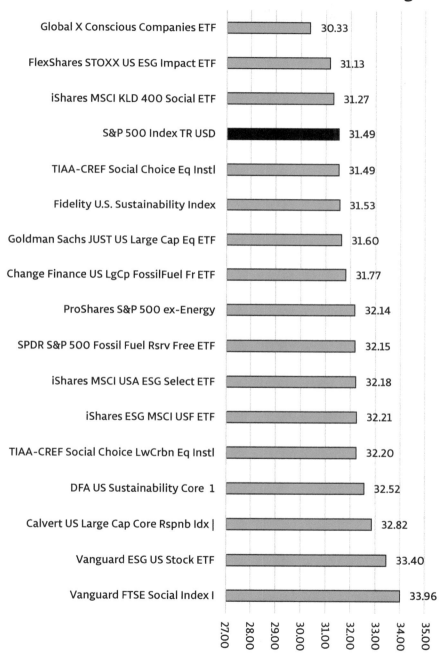

## 2019 Return in Percentage

| Fund | 2019 Return |
|------|-------------|
| Global X Conscious Companies ETF | 30.33 |
| FlexShares STOXX US ESG Impact ETF | 31.13 |
| iShares MSCI KLD 400 Social ETF | 31.27 |
| S&P 500 Index TR USD | 31.49 |
| TIAA-CREF Social Choice Eq Instl | 31.49 |
| Fidelity U.S. Sustainability Index | 31.53 |
| Goldman Sachs JUST US Large Cap Eq ETF | 31.60 |
| Change Finance US LgCp FossilFuel Fr ETF | 31.77 |
| ProShares S&P 500 ex-Energy | 32.14 |
| SPDR S&P 500 Fossil Fuel Rsrv Free ETF | 32.15 |
| iShares MSCI USA ESG Select ETF | 32.18 |
| iShares ESG MSCI USF ETF | 32.21 |
| TIAA-CREF Social Choice LwCrbn Eq Instl | 32.20 |
| DFA US Sustainability Core 1 | 32.52 |
| Calvert US Large Cap Core Rspnb Idx \| | 32.82 |
| Vanguard ESG US Stock ETF | 33.40 |
| Vanguard FTSE Social Index I | 33.96 |

Source: Morningstar Direct, 12/31/2019. Note: Oldest share classes of OE funds shown

**Chapter 14**

# A NEW LEVER

**This chapter was contributed by John Boyer, head of Product Development at EntentVest**

"Give me a lever long enough and a fulcrum on which to place it, and I shall move the world." – Archimedes of Syracuse

A lever is a simple device that can yield profound consequences. The genius mathematician Archimedes first formalized the concept of a lever over 2000 years ago, though levers had been used by humans for several thousand years before that. Mechanically, levers can be constructed to balance forces of different magnitudes, or to amplify either force or speed at the expense of the other. Leverage can be understood in a strategic sense as well, as the ability to use resources to accomplish challenging goals efficiently. The concept of leverage is well known to the financially savvy, as it applies to the use of debt to amplify the magnitude of the value change of an investment. I'd like to extend this analogy to illustrate the potent strength of sustainable investment strategies that thoughtfully and deeply interact with the resources available to them – not only financial capital, but intellectual capital in the form of impact frameworks and policies and physical capital in the form of technology and natural resources. I'd also like to highlight some features of effective sustainable investment strategies, warn against the downsides of ineffective strategies, and suggest some guiding principles in the creation of such strategies.

An attentive reader of this book is now well aware of a broad set of industries that affect, or are affected by, climate change and the importance and urgency of investing with sensitivity to climate change. Adapting to climate change, especially mitigating potential negative impacts of climate change, is a lofty goal, and for this goal to succeed, our planet needs decisive action.

Global development can only happen as sustainably as we can power it. At the dawn of the 21st century, we have begun to understand as a species the magnitude of our impact on the environment and climate, and have realized that, if unchecked, that impact may soon be irreversible. Groundbreaking scientific advancements and climate modeling techniques have only begun to quantify what we know viscerally to be true: that the rate at which we are polluting our atmosphere is unsustainable. Climate change is among the most pressing threats to our species. As humanity starts to venture beyond the confines of our home world — for exploration today, and with hope for future colonization — the imperative is clear: our best solutions are Earth-focused. Perhaps foreign worlds will one day offer a habitat for the adventurous few, but in a practical and immediate sense we are a 1-planet species: there is no Planet B.

Rising temperatures in the atmosphere and the oceans strain our existing infrastructure to the limit, impeding our efforts to feed and house a growing global population. We are already grappling with the dire effects of increased storm activity; in 2017 alone, economic loss due to natural disasters was estimated at $306 billion dollars. In the last decade, on average, 26.4 million people have been displaced from their homes each year (almost 1 person per second) by disasters brought on by natural hazards. The reality is that unless we immediately and drastically alter the way we produce and consume the energy we need, we may already be too late to prevent long term or even permanent damage from taking place.

A sustainable investment strategy has two core goals: to deploy capital to produce a sustainable impact and a financial return. The latter without the former is business as usual; the former without the latter is philanthropy — a noble undertaking, but outside the scope of this inquiry. The first question one might ask relates to the tradeoffs here: is the goal to produce the best-performing portfolio that meets some baseline of sustainable performance, or the most impactful investment strategy that meets some baseline level of risk-return expectation? I will argue for the latter as a more compelling option.

First, any experienced investor understands the difficulty of beating the market. This is true whether an investment strategy is sustainable or not. The risk of the performance first, sustainability second paradigm is that a portfolio that sacrifices sustainability in pursuit of performance risks achieving neither if markets do not cooperate. There are philosophical issues with this framework as well, when the investor's mindset becomes one where sustainability is a tradeoff versus financial performance. As discussed in Chapter 13, the opposite is more likely true, that more sustainable companies generally tend to outperform financially over the long term.

It's helpful to understand how existing asset managers have frequently approached the task of constructing a sustainable investment strategy. A common approach is as follows: start with an existing non-sustainable portfolio model, screen each of the constituents against some ESG data set, drop the most egregious violators of sustainable norms, and brand your new creation as "socially responsible." Or, even without dropping any constituents, use that same ESG data to re-weight the constituents, increasing the weight of companies with high scores and reducing that of lower-performing companies. Et viola, a sustainable portfolio!

Except – is it? Anyone familiar with the 1984 cult classic mockumentary This Is Spinal Tap will fondly remember a scene in which the protagonist, lead guitarist Nigel Tufnel, proudly shows off an amplifier where the volume knob goes to 11. Pressed to explain the benefit of such a modification, his response is simply, "It's one louder." I believe many existing ESG-branded portfolio offerings suffer from this same fallacious reasoning. By far the easiest way to interpret ESG data is to look at the top-line score for each company, often presented as a number between 0 and 100. Calculating a weighted average of the scores of each underlying security, the asset manager victoriously declares, "The benchmark index scores a 55 on ESG, while our sustainable version scores a 62!" Such strategies are indeed "one louder."

The shortcoming of this approach lies in the nature of the score itself. A company is far too complex an entity to be described by a single number. Corporations today are massive, global entities, which interact with their environment — the biosphere as well as the social/societal/cultural/economic environment in which they exist — in a complex fashion. Accordingly, a company can achieve a certain score on any ESG survey in a myriad of ways. Again, tradeoffs dominate the discussion: a company with great environmental policies but poor governance structures might score exactly the same as a competitor with a horrible pollution record yet strong internal policies. Even the most casual analysis reveals immediately that these two companies are not equivalent in a sustainable sense. Despite this, many of the "sustainable" ETFs seem to fall into this category. I don't believe there is malice involved here, but rather that Occam's razor applies: that is, the best answer for why this has happened is simply that this approach is easier.

By relying on the top-line ESG score, we lose much of the value of the underlying analysis, and when this practice is iterated at the top level, a portfolio can easily be produced that achieves a higher ESG score without delivering any benefits of sustainability. The only objective actually delivered here is an increase in an abstract metric, as if the world's problems could be solved by getting this number to 100 — as if, by force of imagination, the world's oceans would stop rising, the atmosphere would magically free itself of excess greenhouse gases, and the glaciers would freeze in their tracks. In reality, the abstractness of this metric means that the portfolio is really pulling in different directions, trying to optimize all outcomes at once and really achieving none. Imagine pulling on a lever from all sides at once: you don't get much work done.

Proponents of this type of sustainable investment might next argue that, at worst, it's harmless, and at best, anything is better than nothing, right? Not so. A risk of this approach is that it might convince the target audience that something meaningful is actually happening, and thus prevent more meaningful change from occurring down the line. Former BlackRock Chief Investment Officer for Sustainable Investing, Tariq Fancy expressed his concern eloquently, saying in an interview last month: "I came to the conclusion that society is a cancer patient and climate change is growing like a cancer. We're selling wheatgrass to the cancer patient. It's well marketed, but there is no evidence it is going to help stop the spread. After I left that I saw the marketing reach level that implied that by doing this you can fight climate change and impact social causes, it was a rush to gather assets. I started to wonder whether it was actually harmful because it had a placebo effect. If you go back to the wheatgrass analogy, it's kind of like the wheatgrass is being marketed in a way that it is actually delaying the patient from starting chemo. At that point you can't make the argument that it's harmless, it is delaying the reforms we need."

The fatal flaw of these investment strategies is that they fail to set a meaningful goal. How can we do better? The best way to start constructing a sustainable investment strategy is to set an objective aligned with global initiatives and frameworks that is specific, measurable, and data-driven. Choosing an organizing framework like the UN's Principles for Responsible Investing (PRI) is a good place to start. The PRI lists 6 guiding principles for signatories:

**Principle 1:** We will incorporate ESG issues into investment analysis and decision-making processes.

**Principle 2:** We will be active owners and incorporate ESG issues into our ownership policies and practices.

**Principle 3:** We will seek appropriate disclosure on ESG issues by the entities in which we invest.

**Principle 4:** We will promote acceptance and implementation of the Principles within the investment industry.

**Principle 5:** We will work together to enhance our effectiveness in implementing the Principles.

**Principle 6:** We will each report on our activities and progress towards implementing the Principles.

In furtherance of this effort, the PRI publishes research and literature on sustainability issues and offers a set of tools to help investors with reporting their investment practices. There are other frameworks that share a similar intention: to provide a common front for investors to act as a unified force for sustainability. Without a true north star, it is difficult to determine whether an investment strategy is effective in making an impact.

Many existing sustainable investment strategies share another shortcoming. For example, say you'd like to construct a low-carbon portfolio: how can you go about this? Often, the answer is to start with a market-tracking index and then cut out the Energy sector entirely, as this sector contains many of the companies with the highest carbon footprint. Once you run the numbers and adjust the weights, you can achieve a portfolio with a high degree of tracking accuracy to the initial index, with a fraction of the carbon footprint. However, it's premature to declare victory — why?

One reason these types of strategies often fail to meaningfully affect their impact targets is because they avoid holding any stake in the exact types of companies that have the biggest role to play in hitting those targets. Consider how concentrated some of the effects of climate change are. In July of 2017, the

Carbon Disclosure Project found that the top 100 polluters are responsible for a whopping 71% of historical greenhouse gas emissions. Since 1988, more than half of global industrial greenhouse gases can be traced to just 25 corporate and state producers! Clearly, sustainable investment strategies need to interface with these industries, even these exact companies, if they have any hope of long-term success. A lever is only an effective mechanism when it is actually applied against a load.

One way to meaningfully interface with industries that are not traditionally green is to find companies within those industries who are top performers. These will be the companies who are performing best on relevant metrics — and if you've picked an investment goal carefully, you'll know exactly what these metrics should be! While the specific metrics will differ for every investment strategy, frameworks like the United Nation's (UN) PRI will help to define some key indicators.

Another major benefit of equity ownership is the ability to vote on the slate of issues that come up at shareholder meetings. This is a direct channel to company leadership, and investors who vote according to sustainable principles can put direct pressure on those leaders to act. Even if changes aren't immediate, shareholder pressure can result in increased transparency regarding how a company is actually thinking about the relevant sustainability issues facing it. For example, is there transparency into the initiatives within a company? Are these initiatives just refreshed values statements or do they have actionable events and timelines? Are the timelines realistic? Has past performance been consistent with these goals, and is the company on track to meet their stated goals? If a company has reached its goal ahead of time, were the goals ambitious enough? Has the company re-upped its efforts? All of these questions can help differentiate companies who have serious intentions to improve their sustainability from those who just want to talk the talk.

Thoughtful sustainable investment has the potential to be a powerful lever in effecting changes we wish to see in our world and can help accomplish our sustainability goals more efficiently than we might through other methods alone. Effective sustainable investment strategies share a core set of features. They identify a meaningful set of goals which they wish to help achieve, they identify measurable targets within these goals, and they invest in companies that have a meaningful stake in these goals and are outperforming their peers on achieving them.

# Chapter 15

# ADAPTING TO A NEW CLIMATE

Climate change is real. We know that some places are getting warmer while others are getting colder. There have been unprecedented superstorms, droughts, and sea level rise. Fresh water has become scarcer, causing food to become more difficult and more costly to produce. Some of the world's population will have to relocate. The horse is out of the barn with climate change, but we still have a chance to find success by adapting and working to mitigate the negative effects of climate change. The COVID-19 pandemic has proven that even short-term behavior changes, such as driving less and reducing industrial pollution, have cleaned the air and reduced the negative impacts on climate change in a measurable way. While mitigation strategies aimed at lessening the impact of climate change through better practices, better use of resources, and smarter living and are essential, I believe it's too little, too late. However, I also believe that our new normal will present both opportunities and threats as our changing climate unfolds. It is my contention that mitigation and adaptation can work together.

While I now I live in the suburbs of San Francisco, I grew up in the city. I remember when the average temperature was 59 °F, even in the summer, because of the marine layer of fog associated with moist warm air being dragged over a cold current ocean current. I fondly remember those days of cool temperatures allowing for aggressive sports without getting too hot. Later, I moved south to Redwood City, famous for a sign welcoming newcomers that says, "Climate Best by Government Test." The slogan was derived from a report from the U.S. and German governments prior to World War I, in which Redwood City tied with the Canary Islands and North Africa's Mediterranean Coast for best climate. When I moved to Redwood City in the early 1990s, the hot days of summer rarely exceeded the high 80s F; now, it's not unusual to see day after day of temperatures in the 100s F. The California droughts have chronically made it difficult to maintain landscaping, keep our cars clean, and allow for other non-essential water usage. Climate change touches nearly all of us in some way; we must adapt. And we should start thinking about adaptation from a personal perspective now before it's too late.

Adapting should lead us to think about where to live through the lens of the changing climate. Humanity has moved around since the beginning of time. In fact, staying in one place over very long periods of time could be considered the outlier from a historical vantage point. People have always been drawn to coastal areas for moderate climates, easy access to water, and an enhanced ability to travel and engage in commerce. But retiring to coastal southern Florida may not be a great idea based on models of rising sea level, as even the more modest scenarios show much of coastal Southern Florida under water; Louisiana has similar projections, especially around New Orleans. Justin Nobel, a writer who covers issues of science and environment, says the five best places to live in 2100 will be Nuuk, Greenland; Egvekinot, Siberia; Bangor, Maine; Buffalo, New York; Iqaluit, Nunavut. "No floods, famine, or war. And the people are nice." (Nobel, 2018)

As we've seen, climate change causes ripple effects throughout every facet of our lives, from nutrition and water to technology and travel — every aspect of where and how we live on the planet. We cannot ignore the connection between climate change and public health. As we shift our locations and behaviors to adapt to climate change, aspects of our collective public health are affected, whether it's anxiety or PTSD associated with drought, flooding, or forced relocation or reduced food and water supplies caused by shifting pest habitats, drought, or increasingly dense populations. We've seen that there are actions we can take and changes we can make to adapt to climate-induced realities and, hopefully, find new kinds of success. And there are ways that investing in new technologies can help mitigate the public health effects of climate change. The pressure on our water supply can be eased through development and expanded use of filtration and desalination technologies, as well as through conservation technologies and techniques such as smart irrigation and dry farming. Shifting our construction technologies and practices by using locally sourced, sustainable material and constructing shelters that require less climate control to be habitable can make housing more affordable and easier to construct while mitigating negative environmental effects. Similarly, investing in renewable energy can both reduce the negative climatic effects of using fossil fuels and expand the geographies in which we can comfortably and safely live. And all of these shifts in attention and behavior can help us be more informed, so we can lessen the mental health impacts that can occur when we're forced to make large changes in behavior and lifestyle quickly or under duress.

We also saw that there are some areas where it is difficult to predict what disruptions will happen, or even if they will be affected.

Defense technology advancement is a big economic player but fraught with danger. We have the ability to make conventional, nuclear, and biological weapons better, cheaper, and faster. The United States has become more and more divided. After the 2016 and 2020 presidential elections, some think the U.S. is on the verge of a civil war. The world as a whole is becoming more tribal and tumultuous, creating even greater conflict over limited resources. Managing conflict could be one of our greatest challenges moving forward.

There is an argument that interplanetary space travel to settle other inhabitable planets is crucial to human survival as a species. Neil DeGrasse Tyson, noted physicist, says that it's more economical and practical to spend resources on fixing the planet Earth versus leaving it. (Hoare, 2019)Gene therapy, while promising in many areas, presents major ethical problems. The ethics of cloning or replicating humans, harvesting organs, and genetically manipulating offspring, while certainly technologically feasible, is controversial at the very least.

## Consider the environment in decisions large and small

Investing in a new climate is not just about your investment portfolio, it's about lots of small decisions. If we lower our carbon footprint and live more simply, we can enjoy the side benefit of spending less money. By being more aware of our surroundings and the climate and our relation to it, we can enjoy a better quality of life and become gentler to the environment at the same time. We must do what we can to lessen the negativity of human impact on the global environment. Be thoughtful about where to live, where to retire, and what to consume. Consider how you eat, use water, and travel. While mitigations would be desirable to lessen or eliminate continued negative impacts, humans have found success throughout time by adapting to their changing environments, and I believe that possibility remains open to us today.

My basic philosophy for investment is to diversify, invest in passive index investments that focus on sustainability, emphasize new technology, and keep investing costs low. Let's unpack the list.

1. Diversification is the hallmark of investing. Don't have your all of your eggs in one basket.

2. Invest in passive indexes. Over the long haul, for the average investor, pas-

sive index funds have outperformed actively managed funds, with much lower fees.

3. Sustainable portfolios often outperform non-sustainable portfolios with similar fees. There is a lot of data that supports the concept that sustainable companies that demonstrate principled business practices are more successful over time than companies that do not exhibit sustainable or ethical practices.

4. New technology has always lead markets, from the Industrial Revolution to microchip technology and to present day internet, AI, and cloud technologies. The best way to get the private sector on board with climate change mitigation is through consumer and investor sentiment. Investing in new technologies is a form of environmental adaptation, such as investing in solar panels or electric car tech. These new technologies can improve the world's carbon footprint while creating new and improved profit centers for both investment and employment.

5. Investment fees are important. Over time, investment fees that are even a little bit higher can have a profound impact on investors' returns. An average millennial could easily pay in excess of $500,000 of unnecessary expenses over their career by not paying attention to investment management fees.

The world is changing; it always has. We can adapt or pay the consequences. We as individuals and as a group must accept the reality of a new physical and economic climate and modify both our thinking and our actions to thrive. Investing in sustainable companies rewards sustainable behavior in a substantial way. The call to action is simple. Invest sustainably to mitigate adverse climate change and make money at the same time. Improve the world and profit from it.

# Appendix:

## Investment Options for Now

The appendix lists companies and ETF investments separated to follow each chapter. These investments are relatively inexpensive and a robust way to get started in investing in a new climate. These ETFs are just examples: the options will change over time as new ETFs become available and others disappear or are merged into other funds. This is just a start to building a portfolio to invest in a new climate.

**Chapter 3: Food**
Mondelez International
Coca Cola
Tyson Foods (TSN)
Mondelez International (MDLZ)
Kellogg Company (K)
Pepsico (PEP)
Sanderson Farms (SAFM)
Conagra (CAG)
Constellation Brands (STZ)
Beyond Meat (BYND)
Agilent Technologies (A)
Archer Daniels Midland (ADM)
Plant Based
Else Nutrition Holdings, Inc (BABYF)
Burcon NutraScience Corp. (BUROF)
Invesco Dynamic Food & Beverage ETF (NYSEARCA:PBJ)
First Trust Nasdaq Food & Beverage ETF (NASDAQ:FTXG)
Invesco Dynamic Leisure and Entertainment ETF (NYSEARCA:PEJ)
Fidelity MSCI Consumer Staples ETF (NYSEARCA:FSTA)
Invesco S&P SmallCap Consumer Staples ETF (NASDAQ:PSCC)
VanEck Vectors Agribusiness ETF (NYSEARCA:MOO)
Invesco DB Agriculture Fund (NYSEARCA:DBA)

**Chapter 4: Water**
Invesco Water Resources ETF (PHO)
Invesco S&P Global Water Index ETF (CGW)
First Trust ISE Water Index ETF (FIW)

**Chapter 5: Transportation**
The main e-bike companies include:
A2B Bicycles (part of the Hero Eco Group), United Kingdom
Benno Bikes, United States
Cytronex, United Kingdom
Beistegui Hermanos, Spain
eROCKIT, Germany
Gocycle, United KingdomGi Fly Bike, United States
Italjet, Italy
Pedego Electric Bikes, United States
Powabyke, United Kingdom
Riese & Müller, Germany
Revelo Bikes, Canada
Solex, France
Specialized Bicycle Components, United States
Superpedestrian, United States
Tidalforce Electric Bicycle by Wavecrest, United States (defunct)
Wing Bikes, United States

iShares Transportation ETF (IYT)
SPDR S&P Transportation ETF (XTN)
Direxion Daily Transportation ETF (TPOR)
SPDR S&P Smart Mobility ETF (HAIL)
Travel Tech ETF (AWAY)
First Trust NASDAQ Transportation ETF (FTXR)

**Chapter 6: Aviation and Space**
Global Jets (JETS)
iShares Transportation Average (IYT)
SPDR S&P Transportation (XNT)
iShares U.S. Aerospace & Defense (ITA)
Procure Space ETF (UFO)
SPDR S&P Kensho Final Frontiers (ROKT)

## Chapter 7: Health Care and Medical Technology

iShares Evolved US Innovative Health Care (IEIH)

First Trust AMEX Biotechnology (FBT)

Principal Healthcare Innovators Index (BTEC)

SPDR S&P Biotech (XBI)

Global X MSCI China Health Care (CHIH)

Invesco Dynamic Biotech and Genome (PBE)

Virtus Life Sciences Biotech Clinical Trials (BBC)

Virtus Life Sciences Biotech Production (BBP)

iShares US Medical Devices (IHI)Vanguard HealthCare Index (VHT)

## Chapter 8: Renewable Energy

Invesco Solar ETF (TAN)

iShares Global Clean Energy ETF (ICLN)

NextEra Energy ETF (NEE)

Brookfield Renewable Partners ETF (BEP)

ALPS Clean Energy ETF (ACES)

## Chapter 9: Computer Technology

S&P Software & Services (XSW)

MSCI Information Technology (FTEC)

iShares Expanded Tech Sector (IGM)

Vanguard Information Technology (VGT)

Technology Select Sector SPDR (XLK)

Quantum Computing and Machine Learning (QTUM)

First Trust Cloud Computing (SKYY)

Wisdom Tree Cloud Computing (WCLD)

GlobalX Cloud Computing (CLOU)

## Chapter 10: Real Estate and Construction

Infrastructure (PAVE)

Infrastructure (IGF)

Global Infrastructure (NFRA)

iShares Global Real Estate (REET)

iShares US Home Construction (ITB)

SPDR S&P Homebuilders (XHB)

**Chapter 11: Artificial Intelligence and Robotics**

Global X Robotics & Artificial Intelligence (BOTZ)

ROBO Global Robotics and Automation Index (ROBO)

ARK Autonomous Technology & Robotics (ARKQ)

Shares Robotics and Artificial Intelligence Multisector (RBO)

AI Powered Equity (AIEQ)

iShares Evolved U.S. Technology (ETC )

# References

**Chapter 1**

Mark, J. J. (2018, March 28). **Fertile Crescent.** *Ancient History Encyclopedia.* Retrieved from https://www.ancient.eu/Fertile_Crescent/

James E. Hansen and Makiko Sato — July 2011

Brantley, Steve and Bobbie Myers, "Mount St. Helens—From the 1980 Eruption to 2000." USGS.gov. March 1, 2005. Accessed: November 21, 2010.

Decker, Barbara and Robert Decker. 1998. Volcanoes 3rd ed. New York: W.H. Freeman and Company.

Francis, Peter. 1993. Volcanoes, A Planetary Perspective, New York: Oxford University Press.

Gabbatt, Adam, "Volcanic Ash Cloud Could Cost European Business Up To 2.5billion Euros, Says EU." Guardian.co.uk. April 27, 2010. Accessed: October 31, 2010.

Scarth, Alwyn and Jean-Claude Tanguy. 2001 Volcanoes of Europe, New York: Oxford University Press.

Sigurdsson, Haraldur. 1999. Melting the Earth, The History of Ideas on Volcanic Eruptions, New York: Oxford University Press.

Tilling, Topinka, and Swanson, "Eruptions of Mount St. Helens: Past, Present, and Future: USGS Special Interest Publication." USGS.gov. July 2, 2005. Accessed: November 21, 2010.

"1985: Volcano Kills Thousands in Colombia." BBC.co.uk. November 13, 1985. Accessed: October 31, 2010.

**Chapter 3**

Malhi, Y., Roberts, J. T., Betts, R. A., Killeen, T. J., Li, W., & Nobre, C. A. (2008). Climate Change, Deforestation, and the Fate of the Amazon. Science, 319(5860), 169-172. doi:10.1126/science.1146961

Nassos Stylianou, C. (2019, August 09). Climate change food calculator: What's your diet's carbon footprint? Retrieved March 15, 2021, from https://www.bbc.com/news/science-environment-46459714

Lagomarsino, V., & Senft, R. (2019, October 04). Hydroponics: The power of water to grow food. Retrieved January 15, 2021, from http://sitn.hms.harvard.edu/flash/2019/hydroponics-the-power-of-water-to-grow-food/

Mekonnen, M.M. and Hoekstra, A.Y. (2010) The green, blue and grey water footprint of farm animals and animal products, Value of Water Research Report Series No. 48, UNESCO-IHE, Delft, the Netherlands.

**Chapter 4**

Bureau, U. (2019). Census.gov. Retrieved September 16, 2020, from http://www.census.gov/

Rodwan, J. G., Jr. (2020). Bottled Water 2019:Slower but Notable Growth, US and International Developments and Statistics. Bottled Water.org, Jul/Aug.

**Chapter 5**

Notteboom, T. and P. Carriou (2009) "Fuel surcharge practices of container shipping lines: Is it about cost recovery or revenue making?". Proceedings of the 2009 International Association of Maritime Economists (IAME) Conference, June, Copenhagen, Denmark.

**Chapter 10**

Amadeo, K. (2020, December 01). Why Buying a Home Helps Build the Nation. Retrieved January 16, 2021, from https://www.thebalance.com/how-does-real-estate-affect-the-u-s-economy-3306018

**Chapter 12**

Chen, J. (2020, August 28). Impact Investing Definition. Retrieved January 17, 2021, from https://www.investopedia.com/terms/i/impact-investing.asp

**Chapter 13**

Huber, B., & Comstock, M. (2017, July 27). ESG Reports and Ratings: What They Are, Why They Matter. Retrieved September 29, 2020, from https://corpgov.law.harvard.edu/2017/07/27/esg-reports-and-ratings-what-they-are-why-they-matter/

Jones, Jeffrey. "Opinion: ESG Investments May Not Be the 'Crowded Trade' Some Have Feared." The Globe and Mail, 15 Apr. 2021, www.theglobeandmail.com/business/commentary/article-esg-investments-may-not-be-the-crowded-trade-some-have-feared/.

**Chapter 15**

Nobel, J. (2018, July 30). The 5 Best Places to Live in 2100. OneZero. Retrieved 2020, from https://onezero.medium.com/the-5-best-places-to-live-in-2100-e4c360ce3a27

Hoare, C. (2019, December 20). Climate change: DeGrasse Tyson's genius plan to 'turn Earth back' as New York under threat. Retrieved January 17, 2021, from https://www.express.co.uk/news/science/1219538/climate-change-neil-degrasse-tyson-plan-save-earth-new-york-city-flooding-spt

### Chapter opener photo credits:

Chapter 1: Earth, NASA
Chapter 2: © Radomír Režný | Dreamstime
Chapter 3: © Valio84sl | Dreamstime
Chapter 4: By Jag_cz | Shutterstock
Chapter 5: By 10incheslab | Shutterstock
Chapter 6: © Snicol24 | Dreamstime
Chapter 7: © Andranik Hakobyan | Dreamstime
Chapter 8: © Joseph Golby | Dreamstime
Chapter 9: © Olga Sapegina | Dreamstime
Chapter 10: © Naropano | Dreamstime
Chapter 11: © Sompong Sriphet | Dreamstime
Chapter 12: By Foryoui3 | Shutterstock
Chapter 13: © Kowit Lanchu | Dreamstime
Chapter 14: © Pachwelna | Dreamstime
Chapter 15: © Denis Burdin | Dreamstime

# About the Author

Scott Schwartz has been an investment advisor and financial analyst for over 25 years. He has an undergraduate degree in geography with an emphasis in weather and climate; and he holds a master's in business with an emphasis on finance. Besides having a professional and academic focus on investing, he has been an airline pilot, global explorer, and a lifelong student of weather and climate.

Scott is a Certified Financial Planner™, Certified Investment Management Analyst®, and Accredited Investment Fiduciary® and is presently the Founder-CEO of EntentVest Inc., a Palo Alto based sustainable investment advisory firm. He lives with his family in the San Francisco Bay Area.

# Acknowledgments

I would like to thank my wife Ann for her advice and her unwavering help in getting this book done. I also want to thank my adult children Jessica and Danielle for their intellectual and moral support.

Brooke King has been a great editor and kept me on track to complete this book. John Boyer contributed extensively to the sustainable investing discussion, a subject on which I value his knowledge. Special thanks to Jill Turney for her assistance with graphics and layout.

Besides so many that have helped me along the way, I would like to call out Leonardo Assis, PhD for his help in the areas of quantum computing and machine learning; Marina Oster, PhD for her help and insights into environmental science; Michael Marquez for his environmental research; Peter Craddock, CEO of the venture capital firm Shoreline Capital; Gail Yoshida for her insights into nutrition science; Roland Haga for his contribution to civil engineering and modern construction techniques; Ralph Kruger, CEO of Passage to India Hedge Fund, for his global investment perspective; Michael Day for his stunning photos; and a special thank you to fellow author Nancy Rose for her support, guidance, and inspiration.

www.ententvest.com

CPSIA information can be obtained
at www.ICGtesting.com
Printed in the USA
BVHW090008280721
613005BV00023B/644

9 781737 332909